A NATURALIST'S GUIDE TO THE

REPTILES

OF

SRI LANKA

A NATURALIST'S GUIDE TO THE

REPTILES
OF
SRI LANKA

Anslem de Silva & Kanishka Ukuwela

JOHN BEAUFOY PUBLISHING

First published in the United Kingdom in 2017 by John Beaufoy Publishing Ltd
11 Blenheim Court, 316 Woodstock Road, Oxford OX2 7NS, England
www.johnbeaufoy.com

Photo captions and credits
Front cover: *main* Sri Lanka Green Pit Viper © Anslem de Silva; *bottom left* Lyre Head Lizard ©
Kanishka Ukuwela; *bottom centre* Common Lanka Skink; *bottom right* Star Tortoise all © Anslem de
Silva; **Back cover:** Lyre Head Lizard © Anslem de Silva; **Title page** Forsten's Cat Snake © Anslem de
Silva; **Contents page:** Black-cheek Lizard © Anslem de Silva

Photo credits:
Main descriptions: Photos are denoted by a page number followed where relevant by t (top), b
(bottom), l (left), r (right) or c (centre).

Mike Anthonisz 167b; **Udaya Chanaka** 28t, b, 29t, b, 30, 43t, b, 46l, 90r, 96, 113t, b, 125b; **Anslem
de Silva** 7; 8t, 8b; 9; 10b; 11t, 11b; 18l, 18c, 18r; 19t, l, c, r; 20t, bl, br, 21t, br; 22t, b; 23t, bl, br, 24t, b,
25t, b, 26b, 27l, 31bl, 32b, 36t, 37bl, 38t, b, 46r, 49t, 50b, 54t, 56t, 57t, 58b, 61b, 62t, 63b, 65b, 67t, b,
69, 76b, 79t, bl, 85t, 89t, 90l, 91b, 92bl, hr, 94, 97t, b, 98t, b, 99b, 102t, b, 103l, r, 104t, b, 107b, 110t,
b, 112t, b, 114t, b, 116t, b, 117bl, br, 118t, b, 119b, 120t, b, 121b, 125t, 126t, b, 127t, 128t, 129t, b,
130t, b, 132b, 133tl, tr, b, 134tr, b, 135t, 136t, b, 137b, 138t, b, 139t, 140t, b, 141t, b, 142tl, tr, b, 143t,
b, 144l, r, 146t, b, 147t, b, 148b, 149t, b, 150t, b, 151t, b, 152t, 153t, b, 154tr, b, 156bl, br, 157t, b, 158l,
159t, b, 160tl, 162b, l, 163t, l, 165t, b, 166r, 167t; **Carl Gans** 95; **Suraj Goonewardena** 45t; 62b; 101;
Thushan Kapurusinhge 26t; **Suranjan Karunarathna** 48, 52t, b, 55, 66; 70t, b, 124b, 127t; **Laxman
Nadaraja** 166l; **Usui Toshikazu** 27r, 45 br, l, 72t, b; **Kanishka Ukuwela** 10t, 16, 21b l, 31t, br, 32t, 33t,
b, 34t, bl, br, 35t, b, 36b, 37t, br, 39t, b, 40t, b, 41t, b, 42t, b, 44t, b, 47t, b, 49b, 50t, 51t, b, 53, 54b,
56b, 57b, 58t, 59, 60t, b, 61t, 63t, 64t, b, 65t, 68t, b, 71t, b, 73t, b, 74t, b, 75t, b, 76t, 77t, b, 78t, b,
79br, 80t, b, 81t, b, 82t, b, 83t, b, 84t, b, 85b, 86, 87t, b, 88t, b, 89b, 91t, 92t, 93t, b, 99t, 100, 105l, r,
106t, bl, br, 107t, 108t, b, 109t, b, 111t, c, b, 115t, b, 117t, 119t, 121t, 122t, b, 123t, b, 124t, 128b, 131,
132t, 134tl, 135b, 137t, 139b, 145t, b, 148t, 152b, 154tl, 155t, b, 156t, 158r, 160tr, b, 161t, b, 162t, br,
163tr, b, 164.

The presentation of material in this publication and the geographical designations employed do not
imply the expression of any opinion whatsoever on the part of the Publisher concerning the legal status
of any country, territory or area, or concerning the delimitation of its frontiers or boundaries.

ISBN 978-1-909612-92-1

Edited by Krystyna Mayer
Designed by Gulmohur Press, New Delhi

Printed and bound in Malaysia by Times Offset (M) Sdn. Bhd.

@ symbol denotes an endemic species

CONTENTS

ABOUT THIS BOOK

This guide, intended for both naturalists and visitors to Sri Lanka, provides an introduction to snakes, lizards, crocodiles, turtles and tortoises found on the island, with quick and easy tips for identification. It features species that are likely to be encountered in the natural and anthropogenic habitats of Sri Lanka, and a few uncommon ones. At the time of writing, 219 reptile species have been recorded from Sri Lanka, and taxonomic work that is in progress is sure to add more species to this impressive list within the next few years.

The species are arranged under their higher taxonomic groups (snakes, lizards, crocodiles and turtles), and further grouped in their respective families, genera and species, which are listed in alphabetical order of their scientific names. Every species covered is accompanied by one or more colour photographs of the animal. Each account includes the vernacular name in English, current scientific name, average total length, vernacular names in Sinhala and Tamil (the latter where available), a description with identification features, distribution (within Sri Lanka and outside the country), and habitat and habit details. The species descriptions provide key identification features of the species, such as the body form and colouration, to help in quick identification of an animal even without capturing it. However, for some skink species (such as *Eutropis* spp. and *Lankascincus* spp.) and geckoes (*Cyrtodactylus* spp. and *Cnemaspis* spp.), features such as scale counts are given since it is impossible to identify these lizards without this information. Note that according to the country's wildlife laws, animals cannot be captured or removed from their natural habitats without official permits, which must be obtained in advance from the Department of Wildlife Conservation. Obviously, care must be taken when photographing venomous snakes, crocodiles and turtles at close quarters.

It is important to note that this work is not comprehensive. Specialist readers may therefore wish to confirm details of identification and general biology with more technical works on the subject (a few are listed on p. 173). The authors may be contacted via email for further information.

Anslem de Silva
Amphibia and Reptile Research Organization of Sri Lanka
15/1 Dolosbage Road
Gampola, Sri Lanka
Email: kalds@sltnet.lk

Kanishka Ukuwela
Senior Lecturer in Zoology
Department of Biological Sciences
Faculty of Applied Sciences
Rajarata University of Sri Lanka
Mihintale, Sri Lanka
Email: kanishkauku@gmail.com

Knuckles mountain range in the Central Highlands is home to many unique, endemic and rare species of snakes and lizards.

PROFILE OF SRI LANKA

Sri Lanka is a humid tropical island in the South Asian region, situated in the Indian Ocean, off the southern tip of peninsular India at latitudes 5°55'-9°51' N and longitudes 79°41'-81°54' E (see also reverse of front cover).

The island is 65,610km² in area, of which 64,740km² is land and the rest inland waters. The physiography of Sri Lanka consists of a central mass known as the Central Highlands. Three distinct peneplains, or erosion levels, are recognized according to elevation and slope features. The lowest, or first, peneplain (sea level to 270m) is the largest, and extends inland from the coast. The second peneplain, or the uplands, extends from 270m to about 910m, and occupies nearly three-tenths of the island. The highlands, or third peneplain, lie at elevations of 910–2,524m.

Sri Lanka is under the influence of the monsoon winds that blow during two distinct periods of the year, producing large quantities of rain seasonally. The south-western region of the island mainly receives rain from the south-western monsoon in June–September. The whole island receives rain from the north-eastern monsoon in November–February. Most activities of reptiles in these areas are synchronized with rainfall through the timing of their activities, especially reproduction, in order to ensure abundant food supply for the young.

There are four major climatic or ecological zones based on rainfall in Sri Lanka. They are the semi-arid zone, dry zone, wet zone and intermediate zone. The semi-arid zone

Scrub jungles of the semi-arid zone are home to the uncommon chameleon and many other reptiles.

Ancient man-made irrigation tanks of the dry zone are home to Mugger Crocodiles, Water Monitors, terrapins and many species of aquatic snake.

receives an annual rainfall of less than 1,250mm per year, while the dry zone receives an annual rainfall of 1,250–1,900mm. The semi-arid and dry zones occupy nearly 60 per cent of the island. About 2 per cent is covered by the wet zone, and it receives an annual rainfall of 2,500–5,000mm per year. The humidity in the wet zone ranges between 75 and 85 per cent. The intermediate zone consists of intermediate climatic conditions between the dry and wet zones. The average annual precipitation of the intermediate zone ranges between 1,900 and 2,500mm.

The vegetation and natural ecosystems of the island are influenced by its geography and climate. The natural ecosystems include forests, grassland, coral reefs, sand dunes, wetlands and mangroves. As a result of the distinct conditions in different ecological zones, different forest types are seen in each of the zones. For example, the lowland wet zone harbours lowland rainforests, while the highland wet zone comprises submontane and montane forests. The vast lowland dry zone is home to dry mixed evergreen forests, while the lowland intermediate zone has moist semi-evergreen forests and the semi-arid zone has thorn forests or scrubland. Much of the natural forests of Sri Lanka have been lost during the last 150 years due to human activities such as agriculture and urbanization. This has resulted in the loss of natural habitat for many forest-dwelling species, though a number of reptiles have managed to carve out niches in some of these altered habitats.

Tropical lowland rainforests are home to many rare and endemic reptiles.

Submontane forests are habitats for many unique species of lizard.

The sandy coastal beaches of the south-west provide nesting grounds for five of the seven marine turtles in the world.

Mangroves are ideal habitats for Saltwater Crocodiles, Water Monitors and brackish water snakes.

Nilgala fire savanna is a grassland habitat where only a few but unique and endemic reptiles can be observed.

FOLKLORE ON REPTILES OF SRI LANKA

For over three millennia, Sri Lanka has been mainly an agricultural society. Pollen evidence from the highest plains, the Horton Plains (2,000–2,300m above sea level) indicates that cereal plant management, together with slash-and-burn techniques and grazing (herding), took place about 17,500 years ago; it is the earliest evidence of agriculture discovered so far, in Sri Lanka as well as in southern Asia. In addition, the majority of the people of the ancient kingdoms in the dry zone in the third century BC to the present have lived in 'tank'-based societies (tanks are man-made reservoirs). For the natives, certain trees and animals were sacred, and were deified or feared as evil; medicinal herbs were also treated with great respect by the ancients. Their beliefs continue to dictate the attitudes of the local inhabitants towards animals and plants today.

Beliefs and traditions about reptiles outnumber those pertaining to other vertebrates of Sri Lanka. However, a fair number of these, for instance the worship of cobras, prevail in India as well and could possibly be of Indian origin. Nevertheless, over the past 3,000 years or more, such traits have taken on an indigenous character. Though the beliefs may appear strange, some of them have helped in the conservation of reptiles. Their analyses reveal that they have stemmed from observations of natural history and external appearances of reptiles. Some of the prevalent common beliefs, cultural practices and proverbs about snakes and other reptiles in Sri Lanka are introduced here.

Among the reptiles of Sri Lanka, snakes both venomous and non-venomous are popular in folklore, due to their mysterious nature and also to the high toxicity of some species. Among the mildly venomous snakes, there are some interesting beliefs about the two vine snakes (*Ahaetulla* spp.) in Sri Lanka. It is believed that the Green Vine Snake is born from Ceylon Caper (**Sin** Wellangiriya *Capparis zeylanica*) fruits. A bite from a Brown Vine Snake is thought to induce paralysis, causing the victim to wither as if struck by lightning, hence the Sinhala name *Henakandaya* (*Hena* = thunder), or 'thunderbolt snake'. Moreover, the locals believe that even the shadow of a Brown Vine Snake is capable of causing envenomation that leads to withering. It is quite obvious that these two beliefs are myths, however, nothing is known about their origins.

Envenoming due to a bite from the Sri Lanka Cat Snake is considered to induce the victim to sleep. It is possible that the inactivity of this nocturnal snake during the day may have lead to its Sinhala name – *Nidi* (sleeping) *mapilla* (cat snake). Perhaps one of the most interesting beliefs about snakes in Sri Lanka is that of the blood-sucking habit of the red-coloured variety of Forsten's Cat Snake, or the *lemapila* (*l* = blood). It is believed that seven of these snakes descend from the roof at night, lowering themselves in succession by hanging on to one another. The one nearest to the sleeping human bites the toe and sucks the blood of the victim. The blood is passed up through all snakes to the seventh snake on the roof. Victims fall into a deep slumber as soon as they are bitten. This process is continued until all seven snakes are sated, leading to the victim's death.

There are interesting beliefs about many of Sri Lanka's non-venomous snakes. One such concerns the Trinket Snake being considered highly venomous in the literature on traditional snake-bite treatment. Envenomation due to the bite of this snake is supposed to turn the whole body of the victim black, including the saliva and urine. The Rat Snake is considered to be the meanest of all snakes in Sri Lankan folklore, and a person bitten by it will not be bitten by other snakes, especially cobras. Many locals believe that the diminutive fossorial blind snakes (Typhlopidae and Gerrhopilidae) creep into the ears of sleeping people; this is suggested by their common Sinhala name, *kanaulla* (*kana* = ear). The two hard, pointed, claw-like spurs (rudimentary hindlimbs) situated on either side of the cloacae of the Indian Python are believed to contain venom.

The highly venomous Spectacled Cobra is feared by locals and also revered by some due to its close association with religion. The natives believe that cobras appear only on *Kemmura* days (that is, on Wednesdays and full-moon days). A fully grown *Kobonaya*, or flying cobra, is said by some to gradually drop its vertebrae from the tail up to the lower end of the hood; once all the vertebrae are lost, two wings appear on either side of the hood, after which the snake flies off to the Himalayas to meditate as a hermit. Ancient texts and sculptures portray cobras carrying a gemstone or precious jewel in the throat, known as the *Nāga Mānikkaya* (*Nāga* = cobra, *Mānikkaya* = gem stone). The cobra regurgitates this gem on certain nights, and flies are attracted to its bright light, which in turn attracts frogs that the cobra feeds on. Different-coloured cobras have also been associated by natives with different social castes.

There are a few beliefs relevant to the marine snakes of Sri Lanka. Some fishermen believe that if you are bitten by a highly venomous 'true' sea snake (Elapidae), the treatment should be carried out while you are at sea, and you should drink sea water three times. Interestingly, some believe that the brackish water-inhabiting Little File Snake spends six months of the year in water and the remaining six on land. In fact, one of its local (Sinhala) names, *'Diya maha goda maha'*, refers to this aquatic and terrestrial habit. It is interesting to note the former generic name of this snake, *Chersydrus* (Greek *chersos* = dry land and *hudor* = water), in reference to this amphibious habit.

There is a considerable amount of folklore about venomous vipers in Sri Lanka. It is said that when Russell's Vipers grow old, they supposedly metamorphose into pythons, hence the Sinhala name *Polon pimbura* (*Polon* = viper, *pimbura* = python). Furthermore, Russell's Viper hatchlings are believed to emerge by tearing open their mother's abdomen, after which they feed on her carcass. Hump-nosed Vipers are believed to originate from refuse or leaf litter, as they are usually found among decaying leaf litter. A bite from the Sri Lanka Green Pit Viper is thought to turn the skin of the victim green. Folklore on the highly venomous Common Krait reflects its dangerous nature. It is believed that if someone is bitten by one at home, they will die on the way to hospital or to the traditional physician, hence the Sinhala name *Magamaruwa*, which literally means 'death on the way'. Similarly, the Tamil name *Yettadi viriyan* means literally 'eight steps', suggesting that the victim can proceed only eight steps after being bitten by a Common Krait.

Among the lizards in Sri Lanka, the agamids are most noticeably widespread, yet there are very few cultural beliefs about them. The elegant Lyre Head Lizard is considered by the natives to be a reincarnation of a pregnant woman who died during pregnancy, and hence is extensively used in witchcraft.

Geckoes play a significant role in the cultural beliefs of Sri Lankan society. They are plentiful in houses and their presence is made obvious through their calls. Most interestingly, the English word gecko originated from the Sinhala name *'gego'* (house lizard). 'Huna kivuwa wage mama kiyanawa' (I am telling you, just as the gecko told) is an often-quoted proverb in the daily lives of Sri Lankans. It is considered unlucky if a gecko calls from the left side if you are about to start a journey, while a call from the right side is considered lucky. Further, if a gecko falls on your head or any other part of your body on a Monday, it may have unfortunate or possibly fatal repercussions, but if it happens on a Sunday, it foretells a victorious result. Some people believe that geckoes change their colour according to different days of the week, and certain geckoes are believed to belong to different castes. Further, geckoes are thought by some to be cousins of crocodiles, while others consider them to be venomous.

The less-frequently seen skinks have also attracted some attention; these non-venomous lizards are traditionally considered to be highly venomous, and those with two tails are supposed to carry a very valuable gem stone in their stomachs. Monitor lizards have also received much attention from the locals. It is considered a bad omen if a Water Monitor enters a property. Though it is not factually correct, the fat of the Water Monitor is considered to be a powerful poison. Contrastingly, the raw tongue of the Land Monitor is inserted in a ripe banana, and fed to children to improve their memories.

The largest reptiles in Sri Lanka, the crocodiles, are not exempt from folklore. It is widely believed that the crocodile mother feeds on her hatchlings. This belief may have arisen due to the ferrying of hatchlings by the mother from their nest to water. Contrastingly, natives also believe that a crocodile will cry if it accidently swallows a hatchling when it gulps down fish. There is a belief that Land Monitors are born from crocodile eggs. In certain parts of the country it is said that crocodiles come ashore, roll mud or clay into a ball, and look into the sun for a while; after this the lump of mud appears like a piece of red flesh, which is promptly swallowed. Stroking or tickling the belly of a crocodile is believed to induce the animal to release ensnared victims. The bed of a crocodile-bite victim should be protected by a cloth canopy that extends to the full length and breadth of the bed, to prevent geckoes defecating on the victim, which would have disastrous effects, including death. A crocodile-attack victim should be taken into their room on a *pavada* (a traditional floor covering of white cloth that is laid out for religious or lay dignitaries to walk on, akin to a red carpet), laid from the main entrance of the house right up to the victim's bedroom. It is sometimes stated that crocodile bites can cause leprosy.

Similar beliefs exist regarding the chelonians of Sri Lanka. According to traditional medicine, marine turtle flesh is supposed to be effective in maintaining and balancing the three humours, *va* (air), *pith* (bile) and *sem* (phlegm), and also in curing rectal and geriatric problems. Many fisher folk test the toxicity of the flesh of the Hawksbill Turtle by throwing a few pieces of its chopped liver to crows. There is a popular belief that a mother marine turtle lies in ambush at sea to devour its progeny. Strangely, terrapins are deified by some and considered lowly by others as they feed on human excreta (one Sinhala name for the hard-shell terrapin is *Goo ibba*, or 'ordure terrapin') and putrefying animal carcasses, and thus are used in witchcraft to cause ill effects on people. One such is the *Mudevi gurukama*, a supernatural practice that is supposed to make homes desolate. People of the dry zone of Sri Lanka believe that if a hard-shelled terrapin comes to agricultural areas, the land will become barren, and if these terrapins congregate in puddles or marsh areas on land, it indicates an impending drought period. Some people believe that the elegant Star Tortoise is supposed to bring ill luck to a household if it wanders onto its property. However, others believe that it is lucky to encounter a tortoise when setting out on a journey. There is also a belief that sitting on a tortoise's shell can cure rectal prolapse.

Snake bite and its management

Sri Lanka is home to several species of venomous snake that may be found in the vicinity of humans. As a result, snake bite is not uncommon and is considered a rural occupational hazard. Here are some precautionary measures.

Of the 80-odd inland snakes, only about 10 species have caused harmful bites to humans. These venomous snakes are relatively easy to identify. Vipers constitute a major portion of the venomous snakes in Sri Lanka. They are relatively slow moving, with stout bodies, narrow necks and broad heads. Their fangs are folded when not in use (solenoglyphous fangs). Cobras are large, heavy-bodied snakes that bear a hood. They have short, fixed fangs (proteroglyphous fangs). Kraits belong to the same family as cobras, but cannot raise their hoods. Sea snakes are large, slender-bodied snakes with flat, rudder-like tails, and are always marine or at least coastal in distribution, though some may travel far up rivers. They too have proteroglyphous fangs. A word of caution: several non-venomous snakes mimic dangerous snakes, and the reverse is also true. Many snakes (such as the cat snakes and keelbacks), which do not bear true venom glands, can nonetheless inflict painful bites, sometimes aided by chewing that may lead to complications.

When in the field, avoid putting your hands inside cracks or holes that might conceal a snake. Wear shoes that completely cover your feet in the forest, and always carry a reliable torch. Headlamps are even better as they keep both hands free. When crossing a fallen tree, look carefully under it on both sides (particularly the blind side). Keep houses and their surroundings free from refuse, as this attracts rats, which in turn attract snakes.

When handling a snake for study or photography, use a restrainer such as a snake hook (typically an 'L'-shaped stick), or better still a snake grabber. The anti-venom serum commonly available in Sri Lanka is made in India and is of the polyvalent type (made from the venom of the 'Big Four'– the Spectacled Cobra, Common Krait, Russell's Viper and Saw-scaled Viper), and is available in most government hospitals in Sri Lanka.

As most bites occur in rural or coastal areas, the person affected needs to be reassured and kept warm; the bitten limb should be immobilized via a crepe (or any other cloth) bandage, and the person must be taken to hospital as quickly as possible, if necessary by being carried. In case of breathing difficulties such as those resulting from envenoming by kraits, the person may require artificial respiration. It is helpful for the treatment if an accurate description of the snake is provided. Do not apply any tourniquets, make any cuts or suck the bite site, as such methods are likely to complicate the treatment as well as the subsequent healing process. Remove any jewellery or wristwatch worn by the victim immediately, since swelling sets in quickly with some snake bites. In the unlikely event of a snake bite, bear in mind that no snake can kill a healthy adult human instantaneously, and if treated appropriately most patients make a complete recovery.

IDENTIFICATION FEATURES OF SNAKES, LIZARDS AND TURTLES

TYPICAL SNAKES

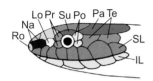

Lateral aspect of head

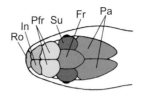

Dorsal aspect of head

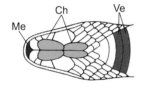

Ventral aspect of head

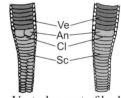

Ventral aspect of body

SKINKS

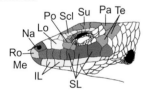

Lateral aspect of head

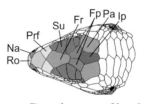

Dorsal aspect of head

TURTLES, TERRAPINS & TORTOISES

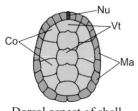

Dorsal aspect of shell

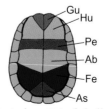

Ventral aspect of shell

▪ GLOSSARY ▪

KEY TO ABBREVIATIONS OF IDENTIFICATION FEATURES

Ab abdominal scute, **An** anal plate, **As** anal scute, **Ch** chin shield, **Cl** cloaca/vent, **Co** costal, **Fe** femoral scute, **Fp** frontoparietal, **Fr** frontal, **Gu** gular scute, **Hu** humeral scute, **IL** infralabial, **In** internasal, **Ip** interparietal, **Lo** loreal, **Ma** marginal scute, **Me** mental, **Na** nasal, **Nu** nuchal, **Pa** parietal, **Pe** pectoral scute, **Po** postocular, **Pr** preocular, **Prf** prefrontal, **Ro** rostral, **Sc** subcaudal, **Scl** supraciliary, **SL** supralabial, **Su** supraocular, **Te** temporal, **Ve** ventral, **Vt** vertebral scute

GLOSSARY

adult Sexually mature individual.

anterior Near the front (towards head).

aquatic Living in water.

arboreal Living in trees or other vegetation away from the ground.

autotomy The shedding of a part of the body when under threat.

canopy Layer of vegetation above the ground, usually comprising tree branches and epiphytes.

clutch Total number of eggs laid by female at a time.

clutch size Number of eggs in a nest.

courtship Behaviour preceding mating.

crepuscular Active during dawn and dusk.

depressed Flattened from top to bottom.

diurnal Active during day.

dorsum Dorsal surface of body, excluding head and tail.

endemic Restricted to a particular region.

femoral pores Pores present on femoral region of some geckos.

fossorial Living underground.

infralabials Scales on lower lip.

ischiadic region Sides of tail region immediately adjacent to cloaca.

keel Narrow prominent ridge.

labial and rostral pits Series of small pits on upper lips of pythons.

lamellae (sg. **lamella**) Pads under digits in lizards (also **scansors**).

litter Detritus of fallen leaves, branches and bark that accumulates on the forest floor.

loreal pit Pit between nostril and eye in pit vipers.

mid-dorsal scales Scales around middle of body.

nocturnal Active during night.

oviparous Egg laying.

ovoviviparous Form of reproduction when eggs develop within body of mother, which does not provide nutrition other than the yolk; eggs hatch inside mother, she later gives birth to live young.

posterior Near back (towards tail).

preanal pores Pores situated before cloaca in geckos.

prefrontals Paired scales on anterior margin of orbit of eye, usually bounded by **frontals**.

proteroglyphous Hollow, short fangs at front of maxilla (upper jaw)

recurved Curved or bent.

reticulate Arranged like a net.

scansors Pads under digits in geckos (also **lamellae**).

scute Horny epidermal shield.

serrated Possessing saw-tooth edge.

subcaudals Scales below tail.

supralabials Scales on upper lip.

tubercle Knob-like projection.

tympanum Eardrum.

ventrals Scales under body, from throat to vent.

vermiculation Pattern consisting of vague, worm-like markings.

vertebral Pertaining to region of backbone.

viviparous Form of reproduction when mother gives birth to live young.

TESTUDINES
The order Testudines comprises reptiles characterized by having bodies enclosed in shells consisting of a dorsal carapace and a ventral plastron. They include marine turtles, freshwater terrapins and land tortoises. The shell is formed through the fusion of ribs. The jaw has tough, horny plates for gripping food. Nine species (five marine turtles, three freshwater terrapins and one tortoise) in six families are known from Sri Lanka. One species is restricted to Sri Lanka, while another has been introduced through the pet trade.

EMYDIDAE (FRESHWATER TERRAPINS)
Emydids are primarily freshwater testudines, though some species inhabit brackish waters and a few are terrestrial. Their sizes are variable (11–60cm in length). In some species the carapace is domed, but most have a low-arching carapace. Ninety-five species in 33 genera are known from North America, northern South America, Europe, north-western Africa and Asia. A single introduced species occurs in Sri Lanka.

Red-eared Slider ▪ *Trachemys scripta* 30cm
(*Sinhala* Rathu-kopul Gal Ibba)

DESCRIPTION Body medium sized. Carapace oval and flattened. Forelimbs and hindlimbs partially webbed, and fingers bear long claws. Carapace has dark green background with light and dark, highly variable markings. Head and limbs dark greyish with yellow stripes. Lateral side of head bears bright red or orange patch behind eyes. DISTRIBUTION Imported to Sri Lanka around 1992 by aquarium traders to be sold for the pet trade. Some individuals escaped or were released into natural waters by pet owners. Currently, there are known established populations in the Western Province. One stray specimen was also seen in the Central Province (Gampola). Native to south-central and eastern USA; subspecies *T. s. elegans* has been introduced to many countries worldwide. Due to its highly competitive nature, considered to be one of the 100 worst invasive species. HABITAT AND HABITS Favours fresh water, inhabiting marshes, lakes, ponds and streams. Fully aquatic species. Feeds on aquatic invertebrates, fish and plants. Female lays 2–30 eggs in a burrow dug on land.

Dorso-lateral   *Ventral*   *Juvenile*

Testudinidae (Land Tortoises)

The Testudinidae family comprises the terrestrial testudines. Their carapace length is 12–130cm. They are characterized by a high-domed carapace, pillar-like legs and blunt, heavily scaled, clawed feet. They are mainly herbivorous, but also occasionally scavenge on carrion. Adapted for a life on dry land, many species can survive long periods without water. A few from wetter parts of the range are, however, fond of water. Tortoises are known dispersal agents of forest plants. They are popular in the pet trade, and large numbers are collected illegally from the wild for export. Approximately 50 species in 11 genera are known from North and South America, Europe, Africa, Asia, and the islands of Madagascar, the Galapagos and the Aldabra Atoll. Only one species occurs in Sri Lanka.

Star Tortoise ▪ *Geochelone elegans* 40cm
(*Sinhala* Tharaka or Mevara Ibba; *Tamil* Katupetti, Manchal amai)

DESCRIPTION Body medium in size. Shell hard and dome shaped, with characteristic yellowish-cream and black-streaked markings on carapace and plastron. Two major forms with different shell structure occur in Sri Lanka – one with small, elongate shell, dull in colour, without carapace protuberances (knobs); the other with large, pronounced protuberances on carapace and bright colouration. Plastron of adult male concave; flat in female. **DISTRIBUTION** Mainly restricted to lowlands of dry, intermediate and semi-arid regions, at sea level to 300m above. Extra-limital: eastern Pakistan and north-western and southern India. **HABITAT AND HABITS** Inhabits scrub forests, agricultural fields, grassland, thorn scrub and home gardens. Crepuscular and terrestrial. Feeds on various plants, including grass species, fallen flowers and fruits; also scavenges on animal matter. Feeds voraciously on tender buds and leaves of various types of bean, groundnuts, pumpkin and cucumber species. Female lays 5–10 brittle-shelled eggs in pit that she digs.

Pyramidal shell

Fewer protuberances

Ventral

Hatchling

Geoemydidae (Freshwater Terrapins and Turtles)

This family comprises small to medium-sized (10–80cm long) aquatic turtles and terrapins. They are characterized by webbed toes and hard shells with scutes, and their necks are drawn back vertically. Their carapaces have 24 marginal scutes, and the plastron is composed of 12 scutes. Geoemydids live in freshwater ecosystems, estuarine waters and tropical forests. About 70 species are distributed in the tropics and subtropics of Asia, Europe, North Africa, and Central and South America. Most are herbivores, but some are omnivores or carnivores. Females lay a small number of eggs per clutch. A single species with two subspecies occurs in Sri Lanka.

Sri Lanka Black Terrapin ■ *Melanochelys trijuga* 35cm
(*Sinhala* Gal Ibba; *Tamil* Karuppu Amai)

DESCRIPTION Body medium sized. Carapace hard and bears 3 keels. Two subspecies occur in Sri Lanka: M. *t. thermalis* and M. *t. parkeri*. M. *t. thermalis*: head usually ornamented with orange or red blotches; carapace black; plastron black with yellow border. M. *t. parkeri*: head uniform olive-brown; carapace dull brownish-black. Sutures between carapace bones more or less distinct. Plastron concave in males, flat in females. Toes webbed, with long, pointed nails. **DISTRIBUTION** M. *t. thermalis* widely distributed in all four climatic zones of Sri Lanka up to 700m above sea level. Extra-limital: south-eastern coast of peninsular India. M. *t. parkeri* confined to northern dry lowlands of Sri Lanka. **HABITAT AND HABITS** Found in lakes, marshes, streams and paddy-field drains.

Freshwater-dwelling, nocturnal species that can be seen basking outside water bodies during day. Also encountered far away from water, either crossing roads or foraging. When captured, it struggles and scratches the captor. During droughts, aestivates in forests under leaf litter. Known to eat aquatic invertebrates, fish, grasses, water hyacinth and fruits, and to also scavenge on a variety of dead animals. Oviparous, with female laying average of 5 hard-shelled eggs in pit she digs in close proximity to water.

Ventral

Lateral

Head ornamentation

TRIONYCHIDAE (SOFTSHELL TURTLES AND TERRAPINS)
These medium to large turtles and terrapins are found in freshwater habitats. Some may also occur in brackish water habitats. They are characterized by their carapaces, which lack horny scutes (leading to the name softshells), and are leathery and malleable, especially at the sides. Their feet are webbed and each bears only three claws. They have elongated, soft, snorkel-like nostrils and very long necks. Most species are carnivores, with diets comprising mainly fish, aquatic invertebrates, amphibians, and sometimes birds and small mammals. About 30 species are distributed in Africa, Asia and North America. A single species occurs in Sri Lanka.

Sri Lanka Soft-shell Terrapin ■ *Lissemys ceylonensis* 30cm ℮
(*Sinhala* Kiri Ibba; *Tamil* Pal Aamai)

DESCRIPTION Body medium sized. Shell slightly domed and oval. Shell greyish-green; in some individuals dull brownish-red; unpatterned; plastron off-white with plastral flaps having tinge of pink. Albinos can occasionally be seen. Head long and nostrils set on fleshy, tubate proboscis. Fingers and toes webbed; claws sharp, curved and long. **DISTRIBUTION** Widely distributed in all climatic zones up to 600m above sea level. Endemic to Sri Lanka. **HABITAT AND HABITS** Inhabits lakes, streams and agricultural canals. Aquatic, nocturnal species that can be seen basking during day. May deliver a

painful bite when captured. Feeds on frogs, tadpoles, fish, crustaceans, molluscs, earthworms, insects and water plants, and also scavenges on animal corpses far from water bodies. Oviparous, with female burying 2–8 eggs on bunds in pit about 5–6cm deep. During height of dry season aestivates by burying itself into mud in tanks or sand in dried streams.

Yellow colouration

Lateral

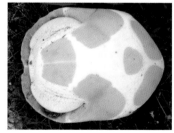
Ventral

CHELONIIDAE (SEA TURTLES)
The Cheloniidae family contains six species of sea turtle in five genera. They have large bodies measuring 70–200cm in length. They are truly marine, with only the females coming ashore to lay eggs. Cheloniids have an oval or heart-shaped shell. Their limbs are modified into flippers for swimming and cannot support their weight on land. Most species are omnivorous and feed on a variety of sponges, cnidarians, molluscs, crustaceans, algae, plants and fish. They occur in tropical and subtropical oceans all over the world.

Loggerhead Sea Turtle ▪ *Caretta caretta* 120cm
(*Sinhala* Olugedi Kasbaeva, Kannadi Kasbaeva; *Tamil* Kadal, Perunthalai Amai)

Dorso-lateral

Dorsal

DESCRIPTION Large body and head. Carapace elongated, with tapering end (heart shaped). Costal scutes 5 pairs. Carapace in various shades of reddish-brown; plastron yellowish-brown or yellowish-orange. **DISTRIBUTION** Known from north-west, south-west, south and south-east of Sri Lanka. Rarest of marine turtles in the Indian Ocean. Extra-limital: Pacific and Atlantic Oceans from Washington, Japan and Kenya, south to Chile, Australia, South Africa, tropical western Africa and Argentina; Caribbean and Mediterranean Seas. **HABITAT AND HABITS** Occurs in shallow coastal seas around Sri Lanka, and known from bays, lagoons and estuaries. Aggressive when disturbed and bites savagely when hauled on board a boat. Feeds on molluscs and crustaceans, and algae. Ovoviviparous, with females laying about 100 eggs on beaches.

Green Turtle ■ *Chelonia mydas* 140cm
(*Sinhala* Gal or Mas Kesbewa; *Tamil* Pachchai Amai)

DESCRIPTION Carapace large and heart shaped; scutes of carapace do not overlap. Carapace olive or brown, usually with dark radiating pattern; plastron pale yellow. **DISTRIBUTION** Common around Sri Lanka. Extra-limital: tropical and subtropical regions of Atlantic, Pacific and Indian Oceans. **HABITAT AND HABITS** Found in shallow waters with seagrass and seaweed. Juveniles are carnivorous, while adults are omnivores, but essentially consume seagrass and seaweed. Known to forage during daylight hours. Adults may also scavenge on refuse. Females lay about 100 soft-shelled, spherical eggs in a pit dug in sand.

Dorso-lateral

Dorsal

Semi-albinistic form

Hawksbill Sea Turtle ▪ *Eretmochelys imbricata* 100cm
(*Sinhala* Pothu or Pana Kasbeva; *Tamil* Ot, Alunk Kadal)

DESCRIPTION Carapace heart shaped; scutes of carapace with 4 pairs of overlapping costals. Upper jaw forwards projecting to form bird-like beak. Adult carapace dark reddish-brown

with variegated yellow markings and slight tinge of green. **DISTRIBUTION** Occurs around Sri Lanka, but rare. Extra-limital: Atlantic (including Mediterranean Sea), Pacific and Indian Oceans. **HABITAT AND HABITS** Homing instincts of species well known. Feeds on sponges and at times algae, gastropods and shellfish. However, being omnivorous it also subsists on seagrass. Clutch size of 60–150 eggs laid in pit dug by female. The species was widely hunted in Sri Lanka from ancient times for its scutes. The flesh can be poisonous, and has caused many human deaths.

Greenish form

Common colour

Olive Ridley Sea Turtle ▪ *Lepidochelys olivacea* 80cm
(*Sinhala* Batu Kasbaeva; *Tamil* Kadai Amai)

DESCRIPTION Smallest and lightest marine turtle. Carapace broad and heart shaped; costals 5–9 pairs. Upper jaw hooked, but lacks ridge. Carapace olive-green or greyish-olive; plastron greenish-yellow; juveniles grey-black dorsally; cream ventrally. **DISTRIBUTION** Common around western, southern and south-eastern parts of Sri Lanka. Extra-limital: tropical and subtropical Americas, the Mediterranean, South Africa and Australia; large-scale nesting occurs in Oman, Mozambique and Florida. One of the largest nesting aggregations (referred to as 'arribadas') occurs in Gahirmatha, Orissa, India, where several hundred thousand turtles congregate to nest. **HABITAT AND HABITS** Feeds on marine vegetation like algae. Also has carnivorous habits and feeds on sponges, molluscs, flat sea urchins, crustaceans and clams, the large jaws being adapted to crush their shells. Nests on beaches with sand bars, sandy islets and dunes with some vegetation. Unlike other local sea turtles, this species does not dig a body pit before nesting. Clutches comprise average of 100 eggs.

Lateral

Dorsal

DERMOCHELYIDAE (LEATHERBACK SEA TURTLE)

The Dermochelyidae family includes just one living species of sea turtle (there are several fossil relatives), the Leatherback Sea Turtle, which is the largest of the sea turtles, reaching 250cm in length, and one of the heaviest living reptiles (800kg). The family is characterized by a carapace that is a composite of osteoderms (small bones) embedded in a leathery skin, hence the common name. Further, these turtles lack any scutes on the carapace, but do have seven longitudinal keels. The species is distributed in tropical, subtropical and temperate oceans all over the world. It frequently wanders into cold Arctic waters, presumably in search of food, and feeds primarily on jellyfish.

Leatherback Sea Turtle ■ *Dermochelys coriacea* 200cm

(*Sinhala* Dhara or Thun Dhara Kasbeva; *Tamil* Dhoni Amai)

DESCRIPTION Shell elongated and tapered towards posterior. Bears 3, 5 or 7 ridges on carapace, and 3 or 5 ridges on plastron. Shell skin-clad in adults, though distinct scales or scale-like structures are found on shells of hatchling. Limbs paddle-like and clawless. Adult may weigh about 900kg. Carapace greyish-black with 5 longitudinal rows of white spots. Plastron creamy-white. **DISTRIBUTION** Locally from Batticaloa downwards to east coast, south-east, south and south-west. Extra-limital: tropical, subtropical and temperate waters of the Atlantic (including the Mediterranean Sea), Pacific and Indian Oceans. **HABITAT AND HABITS** Widely distributed in both warm and cold seas – essentially a turtle of the open ocean. Capable of diving up to 1,200m below the surface in search of prey. Feeds primarily on jellyfish, but in northern temperate waters may occasionally feed on blue-green algae. Females lay 60–150 eggs.

Adult

Hatchling

LIZARDS

Lizards are a broad group of reptiles generally characterized by the presence of limbs and external ear openings, though some species have lost these features due to secondary adaptations. Sri Lanka is home to 104 species of lizard in 10 families (Agamidae, Chameleonidae, Gekkonidae, Lacertidae, Lygosomidae, Mabuyidae, Ristellidae, Scincidae, Sphenomorphidae and Varanidae). Of these, 83 species are endemic to Sri Lanka.

AGAMIDAE (DRAGONS OR DRAGON LIZARDS)

Agamids, or dragon lizards, are small to medium-sized (14-145 cm) diurnal lizards, usually with a crest behind the head that continues to the middle of the back. They are characterized by well-developed limbs, keeled body scales and teeth that are attached to the jawbone. Though the majority are oviparous, a few are ovoviviparous. Most species live on trees and bushes; a few are terrestrial. Nearly all are insectivorous. The family includes more than 450 species distributed in Africa, Asia, Australia and southern Europe. Twenty-one species are known from Sri Lanka, of which 19 are endemic to the island. The species are represented by six genera (*Calotes*, *Ceratophora*, *Cophotis*, *Lyriocephalus*, *Otocryptis* and *Sitana*), of which *Ceratophora*, *Cophotis* and *Lyriocephalus* are restricted to the island.

Green Garden Lizard ▪ *Calotes calotes* 50cm
(*Sinhala* Pala Katussa; *Tamil* Pachai Karata)

DESCRIPTION Large agamid; tail round, long. Head large; cheek swollen in adult males, and crest on head and body distinct in males. Oblique black fold in front of shoulder. Small throat sac. Dorsal scales weakly keeled, pointing backwards and upwards. Dorsum bright green with 5–10 bluish-white or green cross-bars; occasional individuals possess 2 white lines from neck up to mid-tail. Belly pale green. **DISTRIBUTION** Throughout Sri Lanka, but common in lowland and mid-country wet zone. Extra-limital: India. **HABITAT AND HABITS** Frequents grassland, disturbed forests, home gardens, and low vegetation on riverbanks, and around streams and irrigation channels. Diurnal and arboreal. Feeds on insects and occasionally flower petals. Oviparous, laying 6–12 spherical, soft-shelled eggs in a pit. Bites savagely when handled.

Female

Male, lateral

Painted-lip Lizard ■ *Calotes ceylonensis* 30cm ⓔ
(*Sinhala* Thola-visithuru Katussa, Pehethol Katussa)

DESCRIPTION Medium-sized lizard with short dorsal crest. Lateral body scales point

entirely backwards. Two separate, short clusters of spines above tympanum. Males have several colour patterns, such as brownish-black head with distinct white lips, and 3 large, ovate white or red patches along neck. Distinct brownish cross-bars run along dorsal aspect of body, tail and limbs. **DISTRIBUTION** Mainly in lowlands in dry and intermediate zones at up to 500m. Endemic to Sri Lanka. **HABITAT AND HABITS** Confined mainly to edges of monsoonal forests. Diurnal and arboreal, and may descend to the ground for feeding. Found 3–4m above the ground on main tree trunks of large trees in forests and home gardens. When disturbed, runs up along tree trunk to top. Feeds on various insects, such as butterflies and bees, found on tree trunks and flowers. Oviparous, laying about 10 eggs among tree roots.

Male, lateral

Close-up of head

Desilva's Lizard ■ *Calotes deslvai* 30cm ⓔ
(*Sinhala* Desilvage Katussa)

DESCRIPTION Medium-sized agamid with lateral scales on body and tail pointing backwards and downwards. Two separate, short clusters of spines above tympanum. Prominent dorsal crest. Scales on ventral surface of thigh are smooth. Colour variable, from light green with dark stripes to dark brown with light stripes. Lips same colour as head. Gular area has distinct black stripes; shoulder-pit black. **DISTRIBUTION** Restricted to submontane forests of eastern Singharaja and Morningside forest reserves. Endemic to Sri Lanka. **HABITAT AND HABITS** Known only from closed canopy primary forests. Arboreal, and generally found on trees more than 2–5m above ground. Oviparous, with female excavating small pit and laying 2–4 eggs.

Close-up of head

Lateral

Crestless Lizard ■ *Calotes leocephalus* 18cm ⓔ
(*Sinhala* Dethikonda Rahitha Katussa)

DESCRIPTION Body medium sized. Head large, tympanum distinct and no spines above tympanum. Shoulder pit present. Dorsal and lateral body scales uniform; lateral body scales point backwards and downwards; abdominal scales end abruptly in short, sharp points; pectoral scales enlarged. Dorsum bright green to greenish-olive with six 'V'-shaped, light blue or black markings along dorsal surface; 6–7 brown or black cross-bars on limbs. A few cross-bands between eyes. Tail brownish or cream with dark cross-bars. **DISTRIBUTION** Restricted to Central Highlands at more than 1,300m above sea level. Endemic to Sri Lanka. **HABITAT AND HABITS** Found in montane and submontane forests, forest edges, roadsides and tea estates. Diurnal and arboreal. Adults can be seen on large trees. Feeds on insects. Oviparous, laying 3–4 soft-shelled eggs in hole dug in the ground.

Lateral

Whistling Lizard ■ *Calotes liolepis* 48cm ⓔ
(*Sinhala* Sivuruhandalana Katussa)

DESCRIPTION Medium-sized agamid with lateral scales on body and tail pointing backwards. Two separate, short clusters of spines above tympanum; prominent dorsal crest. Scales on ventral surface of thigh keeled. Colour highly variable from light green to dark brown with light stripes. Lips the same colour as head. Gular area has light bands; shoulder pit creamy-white to brown. **DISTRIBUTION** Widely distributed in wet and intermediate zones from sea level up to 800m. Endemic to Sri Lanka. **HABITAT AND HABITS** Forest species that also inhabits monoculture plantations (*Pinus*, *Acacia*, cardamom), home gardens and large trees by roadsides. Slow-moving, diurnal and arboreal, and when disturbed, runs up along tree trunks to 10m or more. Feeds on insects and grubs. Oviparous, with female excavating small pit (50mm deep), and laying 4–6 eggs. Emits low whistle when captured, and bites savagely.

Dark brown

Light brown

Close-up of head

Black-cheek Lizard ■ *Calotes nigrilarbris* 45cm ⓔ
(*Sinhala* Kalu-kopul Katussa, Kandukara Pala-katussa)

DESCRIPTION Large agamid with lateral scales on body and tail pointing backwards or downwards. Males have large head and 2 rows of spines above and below tympanum. Row of dorsal spines runs from neck to hindlimbs. Body generally light green, and males have broad black band on lips running up to end of head. **DISTRIBUTION** Central Highlands above 1,400m. Endemic to Sri Lanka. **HABITAT AND HABITS** Inhabits montane and submontane forests, grassland, cultivation and home gardens. Diurnal and arboreal. Lives on shrubs and trees during day. May descend to the ground to sleep on grass at night. In Horton Plains, can be seen feeding on insects that are attracted to gorse *Ulex europeus* flowers during day. While basking, changes entire body colour to black to absorb heat. Oviparous, with female digging shallow pit and laying 3–5 eggs.

Female, lateral

Male with a black cheek

Pethiyagoda's Crestless Lizard ■ *Calotes pethiyagodai* 17cm ℮
(*Sinhala* Pethiyagodage Dethikonda Rahitha Katussa)

DESCRIPTION Body medium sized. Head large; tympanum distinct; no spines above tympanum. Shoulder pit present. Dorsal and lateral body scales uniform; lateral body scales point backwards and downwards; abdominal scales acuminate; pectoral scales not enlarged. Dorsum bright to dark green with eight 'V'-shaped, light blue or black markings along dorsal surface. Incomplete greenish-brown or black cross-bars on limbs; 2–6 black cross-bands between eyes; tail brownish or cream with 8–10 black cross-bars. **DISTRIBUTION** Restricted to Knuckles mountain range at more than 1,000m above sea level. Endemic to Sri Lanka.

HABITAT AND HABITS Found in submontane forests, forest edges, roadsides and tea estates in Knuckles mountain range. Diurnal and arboreal. Adults can be seen on large trees while juveniles are seen on shrubs. Feeds on insects such as butterflies, moths and dragonflies. Oviparous, laying 3–8 soft-shelled eggs in hole dug in the ground.

Close-up of head

Dorso-lateral

Common Garden Lizard or Oriental Garden Lizard

■ *Calotes versicolor* 50cm
(*Sinhala* Gara Katussa; *Tamil* Karata)

DESCRIPTION Head large. Scales on body point backwards and upwards. No fold or pit in front of shoulders. Two separate sets of spines above tympanum. Colouration variable and also changeable, with head becoming bright red and black patch on throat in displaying males, fading to dull grey at other times; females may become yellow, changing to dull greyish-olive after mating. **DISTRIBUTION** The most abundant and widespread agamid in Sri Lanka, restricted to elevations from sea level to 1,000m above. **HABITAT AND HABITS** Mostly seen in anthropogenic habitats, though some occasionally seen in pristine habitats as well. Diurnal and arboreal. Feeds on insects and other invertebrates. While basking, the lizards try to orient their body surfaces towards sunlight and keep the body colour dark to increase heat absorption. Head bobbing and dewlap expansions are common when another individual is encountered. Oviparous, laying about 12 eggs in a pit.

Male, lateral

Common colouration

Close-up of head

Lyre Head Lizard ■ *Lyriocephalus scutatus* 35cm ⓔ
(*Sinhala* Getahombu Katussa, Karamal Bodiliya)

DESCRIPTION Easily distinguished from other agamids by presence of globular, knob-like rostral appendage. Distinct dorsal superciliary ridge extends above and beyond eye. Distinct row of dorsal serrated scales along spine from neck to tail-tip. Tail flattish. Body colour highly changeable from dark to light green, especially in males. Hatchlings and juveniles grey-brownish. **DISTRIBUTION** Widely distributed in wet and intermediate zones at 30–1,000m above sea level, though habitat is highly fragmented. Endemic to Sri Lanka. **HABITAT AND HABITS** Lives in close-canopy primary and secondary forests with dappled sunlight, and also on trees in humid home gardens. Diurnal, arboreal and slow-moving species, usually encountered on tree trunks about 2m from the ground. Feeds on earthworms, grubs and insects. Oviparous, laying 10–16 eggs in pit. When threatened, displays inside of its bright red mouth.

Female, lateral

Male, lateral

Rhino Horn Lizard ■ *Ceratophora stoddartii* 25cm ⓔ
(*Sinhala* Kagamuva Angkatussa)

DESCRIPTION Distinguished from all other agamids in Sri Lanka by presence of rostral appendage similar to a rhino's horn, which is prominent in males but less prominent in females. Large, irregular scales on body's lateral sides. Body dull greenish-brown; chin whitish. **DISTRIBUTION** Widely distributed in Central Highlands at 1,200–2,300m. Endemic to Sri Lanka. **HABITAT AND HABITS** Occurs in pristine submontane and montane forests, open die-back areas in forests, secondary forests, tea plantations and home gardens in highlands. Diurnal, arboreal and slow moving. During day can be seen on tree trunks 1–2m above the ground. May descend to the ground to feed on caterpillars, earthworms and arthropods. On the ground changes body colours to brown, camouflaging itself in the leaf-litter. Oviparous, with female laying about 5 eggs in pit that she digs.

Female

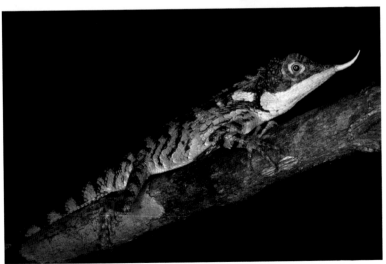

Male, lateral

Leaf-nose Lizard ■ *Ceratophora tennentii* 25cm ⓔ
(*Sinhala* Pethi Angkatussa, Dumbara Angkatussa)

DESCRIPTION Distinguished from all other agamids in Sri Lanka by flat, fleshy, leaf-like rostral appendage that is more prominent in males than females. Body colour highly variable, from greenish to dull greenish-brown. Large, rectangular scales on lateral sides of body. Tail long and cylindrical. **DISTRIBUTION** Confined to moss and lichen-covered forests of Knuckles mountain range at 700–1,700m above sea level. Endemic to Sri Lanka. **HABITAT AND HABITS** Found in submontane and montane forests, but can also be seen in cardamom and pine plantations. Diurnal, arboreal and slow moving. During day may be seen on tree trunks 1–2m above the ground, on low vegetation or on rocks. Feeds on caterpillars, arthropods, grubs and ants. Occasional cannibalism has been noted. Oviparous, laying 3–5 eggs in a pit.

Male, lateral

Female

Male, close-up of head

Rough-horn Lizard ■ *Ceratophora aspera* 10cm ⓔ
(*Sinhala* Ralu Angkatussa, Kuru Katussa)

DESCRIPTION Small agamid. Distinguished from all other agamids of Sri Lanka by numerous spiny projections on pointed rostral appendage, which are more prominent in males than females. Head and body have many spiny tubercles. Limb and tail scales spiny (keeled). Body dull brownish-red with tinge of green. **DISTRIBUTION** Lowland wet zone at 90–900m above sea level. Endemic to Sri Lanka. **HABITAT AND HABITS** Lives on forest floor of well-shaded lowland rainforests. Diurnal, terrestrial and slow-moving agamid that seldom climbs low vegetation. During day may be seen on leaf litter, at bases of moss- and lichen-covered trees, on decaying logs and on small rocks on the ground. Feeds on caterpillars, insects, grubs and ants. Oviparous, laying 2–3 eggs in a pit.

Male, close-up of head

Dorso-lateral

Pygmy Lizard ▪ *Cophotis ceylanica* 15cm ℮
(*Sinhala* Kandukara Kurukatussa, Kuru Bodiliya)

DESCRIPTION Small agamid distinguished from all other agamids of Sri Lanka (except Knuckles Pygmy Lizard, see p. 40) by long snout, indistinct tympanum, prehensile tail and backwards-pointed, large, irregular scales on dorsal aspect of body. Males have small, distinct gular sack. Bears 22–27 sharply pointed, keeled gular scales; 3 nuchal crest spines; 13 dorsal crest spines; and keeled acuminate chest scales. Dull yellowish-green with about 10 brownish-green cross-bars from neck to tail-tip. **DISTRIBUTION** Confined to Central Highlands at 1,500–2,200m above sea level. Endemic to Sri Lanka. **HABITAT AND HABITS** Lives in montane forests, and also anthropogenic habitats like cypress hedges in home gardens and low vegetation in tea plantations. Diurnal, arboreal and slow moving. Can be seen on small twigs in shrubs, and also large trees about 9m above ground level. Feeds on caterpillars, arthropods, grubs and ants. Occasional cannibalism has been noted. Ovoviviparous, giving birth to 2–8 young.

Male, close-up of head

Male, lateral

Knuckles Pygmy Lizard ■ *Cophotis dumbara* 14cm ⓔ
(*Sinhala* Dumbara Kurukatussa, Dumbara Kuru Bodiliya)

DESCRIPTION Small lizard with prehensile tail. Tympanum not very distinct, dorsonuchal crest well developed in adults, gular sac small, lateral body scales large and irregular in shape, and tail prehensile. Colour highly variable and no distinct pattern. However, labial region is lightest, with distinct light bands on tail. Colour can be changed rapidly to blend with surroundings. Distinguished from Pygmy Lizard (see p. 39) by presence of 34–35 smooth, pointed scales on gular sac, 2 nuchal-crest spines and feebly carinate obtuse scales on chest. **DISTRIBUTION** Restricted to Knuckles mountain range

above 1,200m. Endemic to Sri Lanka. **HABITAT AND HABITS** Found on moss- and lichen-covered tree trunks, branches and twigs, in submontane and montane cloud forests and cardamom plantations. Diurnal, arboreal and slow moving. Diet mostly comprises small insects such as moths. Assumed to be ovoviviparous, giving birth to 2–3 young. Critically Endangered.

Male, close-up of head

Lateral

Lowland Kangaroo Lizard ■ *Otocryptis nigrisitigma* 25cm ⓔ
(*Sinhala* Kalulap thali katussa, Yak Katussa)

DESCRIPTION Small agamid with slender body, and long hindlimbs and tail. Five toes on hindlimbs. Males have distinct dewlap with distinct black spot. Body and tail dull brownish-grey. Male's head, neck region and dewlap often become luminescent greenish-blue. Female light brownish-yellow, with head and neck region dull black.
DISTRIBUTION Confined to intermediate zone and dry-zone lowlands from sea level to 600m above. Endemic to Sri Lanka. **HABITAT AND HABITS** Found in dry-zone scrubland, tropical mixed evergreen forests, areas with vegetable cultivation and home gardens. Diurnal, terrestrial, fast-moving agamid that uses bipedal gait when running. Runs for short distance and hides or stops for a few seconds. At night ascend to short vegetation to sleep, hatchlings hide under leaf litter or crevices. Feeds on caterpillars, insects, grubs and small geckoes. Oviparous, with female laying about 3–5 eggs in pit she digs.

Male, close-up of head

Male, lateral

Sri Lanka Kangaroo Lizard ▪ *Otocryptis weigmani* 25cm **e**
(*Sinhala* Rathulap Thali Katussa, Yaka katussa, Pinum katussa)

DESCRIPTION Small agamid with slender body, and long hindlimbs and tail. Five digits on hindlimbs. Male has well-developed dewlap with distinct maroon patch. Body colour to tail tip dull brownish-grey. Male's head, neck region and dewlap often become luminescent greenish-blue. Female light brownish-yellow with dull black head and neck. **DISTRIBUTION** Mainly confined to wet zone from sea level to 1,300m above. Endemic to Sri Lanka. **HABITAT AND HABITS** Found on forest floor, among roots of large trees, on low vegetation and even on boulders along streams. Diurnal, terrestrial, fast-moving agamid that uses bipedal gait when running. Runs for short distance and hides or stops for a few seconds. Feeds on caterpillars, arthropods and grubs. Sleeps on low vegetation at night. Oviparous, with female laying about 3–5 eggs in pit she digs.

Female, lateral

Male, lateral

Bahir's Fan-throat Lizard ■ *Sitana bahiri* 18cm **ⓔ**
(*Sinhala* Bahirge Thali Katussa)

DESCRIPTION Small agamid with slender body, and long hindlimbs and tail. Four toes on each hindlimb. Males have large dewlap (but shorter than that in Devaka's Fan-throat Lizard, see p. 44), with orange patch in middle and blue patch in outer edge. Abdominal scales bluntly pointed. Dorsum brown, with dark brown, black-edged, diamond-shaped marks. Mouth lined with dark blue. Belly cream. **DISTRIBUTION** Dry coastal areas of south-east from sea level to 50m above. Endemic to Sri Lanka. **HABITAT AND HABITS** Found in dry coastal vegetation, sandy areas and scrubland. Fast, bipedal, diurnal lizard of open areas. Capable of running on hindlimbs, with tail raised. Males are highly territorial. Feeds on ground-dwelling insects. Oviparous, laying 4–6 eggs in hole dug in cool areas under shrubs.

Female

Male, lateral

Devaka's Fan-throat Lizard ▪ *Sitana dewakai* 18cm ⓔ
(*Sinhala* Dewakage Thali Katussa)

DESCRIPTION Small agamid with slender body, and long hindlimbs and tail. Four toes on each hindlimb. Male has large dewlap (longer than in Bahir's Fan-throat Lizard, see p. 43), with bright red colour patch in middle and Persian blue patch in outer edge. Abdominal scales are pointed. Dorsum is brown, with dark brown, black-edged, diamond-shaped marks. Mouth lined with dark blue. Belly cream. **DISTRIBUTION** Dry coastal areas of northern and north-western Sri Lanka from sea level to 50m above. Endemic to Sri Lanka. **HABITAT AND HABITS** Found in dry coastal vegetation, sandy areas and scrubland. Fast, bipedal, diurnal lizard from open areas that is capable of running on hindlimbs, with tail raised. Males highly territorial. Feeds on ground-dwelling insects. Oviparous, laying 3–6 eggs in a hole dug in cool areas under shrubs in September–November.

Female

Male, lateral

CHAMAELEONIDAE (CHAMELEONS)
Chameleons are a unique group of lizards that can move their eyes independently and have long, projectile, sticky tongues that they use to capture prey. Their hands and feet are zygodactyl – they have digits in two opposable sets, in which two digits are fused to form the forward set and three digits are fused to form the other. Most are arboreal and the majority have prehensile tails. They are slow-moving lizards that have the ability to change their colours rapidly. About 200 species are known from Africa, Madagascar, southern Spain, the Arabian Peninsula, India, Pakistan and some Indian Ocean islands. Their diversity peaks in Madagascar. Sri Lanka is home to a single species, which is shared with India.

Sri Lankan Chameleon ■ *Chamaeleo zeylanicus* 40cm
(*Sinhala* Bodiliya, Bodilima; *Tamil* Pachai Wona)

DESCRIPTION Large lizard with compressed body, and head with distinct bony casque. Orbit of eye large, and eyeball covered with skin, leaving tiny aperture. Scales on body enlarged and tuberculate. Low, serrated dorsal crest and prehensile tail. Fingers and toes opposable. Colour changeable from green to yellow, and dark spots and stripes extending from edges. **DISTRIBUTION** Semi-arid and dry zones. Extra-limital: India. **HABITAT AND HABITS** Found on low vegetation in scrub jungle, open woodland and occasionally home gardens. Diurnal, arboreal lizard that has remarkable capacity to change body colour rapidly. Diet comprises insects that are captured with the tongue, which can be extended for nearly the length of the body. Oviparous, laying 10–31 leathery textured eggs in a clutch.

Lateral

Zygodactyl hands

Prehensile tail

> ## GEKKONIDAE (TYPICAL OR COSMOPOLITAN GECKOES)
> Gekkonids are small to medium-sized lizards with bodies that are covered with small, non-overlapping scales. They have fixed eyelids and soft, fleshy bodies. They easily lose their tails in defence (autotomy), and the tails are regenerated. Geckoes are mostly oviparous, laying hard-shelled eggs, though a few are viviparous. Arboreal species have structures called lamellae on the fingers that stick to climbing surfaces. More than 1,000 species are known from Africa, Asia, southern Europe and Australia. Forty-four species in seven genera are known from Sri Lanka. Thirty-five species are endemic to Sri Lanka.

Sri Lankan Golden Gecko ■ *Calodactylodes illingworthorum* 15cm 🄴
(*Sinhala* Maha Galhuna)

DESCRIPTION Head, eyes and body large; pupils vertical; tail long; limbs and fingers long. Fingers dorsally bear 2 subtriangular expansions, which form 2 large lamellae. Body colour cream-brown to light yellow, with series of dark brown spots running dorsally from neck; spots turn into bands on tail. Body heavily mottled with small brown spots, except for small region between dark brown spots that run along body. Limbs have dark brown and lighter bands. Belly whitish-pink. Ventral side of head light yellow. **DISTRIBUTION** Restricted to scattered hills of intermediate and dry zones in Uva and Eastern Provinces, at 150–600m above sea level. Endemic to Sri Lanka. **HABITAT AND HABITS** Lives in caves or rocky areas surrounded by scrub trees and monsoon forests in savannah grassland. Nocturnal and rock dwelling. Active at dusk and dawn. Feeds on beetles, moths, grubs and glow-worms. Characteristic call is the loudest among Sri Lankan geckoes. Oviparous, with several females laying eggs at same site in rocky cave surface for several seasons (communal nesting). Eggs attached to cool, dark place on vertical or horizontal surface of rock cave.

Dorsal

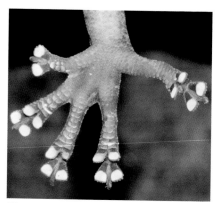

Close-up of hand

Alwis's Day Gecko ■ *Cnemaspis alwisi* 8cm ℮
(*Sinhala* Alwisge Diwasarihuna)

DESCRIPTION Body small and slender. Fingers and toes slender, bent at an angle and tips not expanded. Eyelids prominent, pupils rounded. Dorsal scales on body more or less equal in size; 7–9 femoral pores; subcaudal scales on tail enlarged; abdominal scales smooth; spine-like tubercles on flanks. Head, body and tail dull light brown to greyish with whitish, irregular-sized, circular spots scattered from head to base of tail. A black somewhat rectangular patch present on the neck. Ventral aspect of throat is yellow. Ventral side of tail is greyish white. Three faded white stripes are present on each lower and upper arm in a dark grey background.

DISTRIBUTION Highly scattered on several isolated mountains in wet, intermediate and dry zones. Endemic to Sri Lanka. **HABITAT AND HABITS** Inhabits rocky or boulder-strewn areas in forests and open areas adjacent to forests. Diurnal and fast moving. Feeds on insects. Females known to be communal nesters that lay eggs in crevices in trees and rocks.

Dorso-lateral

Dorsal

47

Gemunu's Day Gecko ◾ *Cnemaspis gemunu* 6.5cm ⓔ
(*Sinhala* Gemunuge Diwasaruhuna)

DESCRIPTION Body small and slender. Fingers slender, bent at an angle, tips not expanded. Eyelids prominent, pupils round. Dorsal scales on body more or less equal in size; abdominal scales smooth; subcaudal scales on tail enlarged; irregular row of spine-like tubercles on flanks; no preanal pores in males; 12–16 femoral pores; 113–115 ventral scales. Body mottled greyish-brown with backwards-pointing, 'V'-shaped marks between axilla and groin, each with thin black margin enclosing triangular area of orangish-brown; dorsal surfaces of limbs mottled with narrow, irregular, dark brown bands. Bold black mid-dorsal line over midpoint of neck. **DISTRIBUTION** Central Highlands of Sri Lanka at more than 1,300m above sea level. Endemic to Sri Lanka. **HABITAT AND HABITS** Lives in montane forests, forest edges and home gardens. Diurnal and arboreal. Forages on moss-covered tree trunks, stone walls and rock surfaces, feeding on insects and spiders. Oviparous, laying 1–2 eggs.

Dorsal

Kandyan Day Gecko ■ *Cnemaspis kandiana* 7cm ⓔ
(*Sinhala* Kandukara Diwasarihuna)

DESCRIPTION Body small and slender. Fingers slender, bent at an angle, tips not expanded. Eyelids prominent, pupils round. Dorsal scales not pointed, keeled, irregular in size; enlarged tubercles on dorsal and dorsolateral sides of body; gular scales keeled; subcaudal scales on median scale row of tail irregular, smooth; 2–4 preanal pores. Dorsum dark brown with 6–7 inverted 'V' markings bordered by lighter region. Some individuals may have yellow dorsal line that runs from neck to base of tail.

DISTRIBUTION Environs of Kandy in Central Highlands at 400–700m above sea level. Species endemic to Sri Lanka. **HABITAT AND HABITS** Found in forests and shady areas in plantations and home gardens. Diurnal, semiarboreal gecko that can be seen on tree trunks and rough walls, and under logs and boulders. Feeds on small insects. Oviparous, with female usually laying 2 globular eggs at a time. Communal nesters in caves, cracks in walls and tree trunks.

Mating pair with yellow dorsal line

Dorso-lateral

Dwarf Day Gecko ■ *Cnemaspis podihuna* 5cm ●
(*Sinhala* Podi Diwasarihuna)

DESCRIPTION Body small and slender. Fingers slender, bent at an angle, tips not expanded. Eyelids prominent, pupils round. Dorsal scales on body more or less equal in size; abdominal scales smooth; 3–4 preanal pores in males; 3–6 femoral pores in males; 113–177 ventral scales; subcaudal scales on median row of tail enlarged, subequal. Dorsum light brown with merging black colour markings on lateral sides of body; striking yellowish or gold stripe on nape. **DISTRIBUTION** Lowlands of intermediate and dry zones. Endemic to Sri Lanka. **HABITAT AND HABITS** Found in forests, scrubland, open areas and home gardens. Diurnal, semiarboreal species that lives on tree trunks. Oviparous, laying 2 eggs in cracks and under tree bark. Several individuals can be seen on trunks of some large trees.

Dark

Light

Jerdon's Day Gecko ■ *Cnemaspis scalpensis* 6cm ⓔ
(*Sinhala* Jerdonge Diwasarihuna)

DESCRIPTION Body small and slender. Fingers slender, bent at an angle, tips not expanded. Eyelids prominents, pupils round. Dorsal body scales more or less equal in size; abdominal scales smooth; conical tubercles in dorsolateral region of body; subcaudal scales in median row of tail enlarged; preanal pores absent in males; 12–16 femoral pores in males; 122–130 ventral scales. Dorsum light brown with about 6 'V' marks from neck to base of tail; nape with distinct black marking; adult males have bright yellow throats. **DISTRIBUTION** Restricted to forests of Kandy area, common in Gannoruwa forest reserve. Endemic to Sri Lanka.

Head, ventral

HABITAT AND HABITS Occurs in forests, forest edges and home gardens. Diurnal and arboreal; can be seen mainly on large tree trunks, and occasionally on large boulders. Fast-moving species. Oviparous, laying 2 eggs under tree bark or in crevices.

Dorso-lateral

Roughbelly Day Gecko ■ *Cnemaspis tropidogaster* 6cm ⓔ
(*Sinhala* Ralodara Diwasarihuna)

DESCRIPTION Body small and slender. Fingers slender, bent at an angle, tips not expanded. Eyelids prominent, pupils round. Abdominal scales keeled; dorsal scales on body heterogeneous; tail base with homogeneous scales; 6 spine-like tubercles on each flank. Dorsum, limbs and tail light brown; oblique dark brown line with dull whitish spots in interorbital area; 'W'-shaped, dark brown patch on neck with median dull white spot; 6 diffused, 'A'-shaped, yellowish markings on trunk dorsum, and reddish-brown band along

tail. **DISTRIBUTION** So far known only from Pilikuttuwa and Maligatenna in Gampaha district, Western Province. Endemic to Sri Lanka. **HABITAT AND HABITS** Occurs in primary and secondary lowland rainforests. Diurnal gecko that can be seen on boulders and cement walls, among leaf litter, under tree bark, and in caves and crevices. Oviparous, laying 2 small, spherical eggs that are attached to a substrate.

Female, dorsal

Male, lateral

Upendra's Day Gecko ■ *Cnemaspis upendrai* 5cm ⊜
(*Sinhala* Upendrage Diwasarihuna)

DESCRIPTION Body small and slender. Fingers slender, bent at an angle, tips not expanded. Eyelids prominent, pupils round. Dorsal body scales dissimilar in size; subcaudal scales in median row of tail enlarged; abdominal scales with single keel; some throat scales with 3 keels; tail-base with tubercles; 112–141 ventral scales. Dorsum dark to light brown with blackish or dark brown blotches; tail reddish or orangish-brown; some individuals may have yellow dorsal stripe from nape to anterior part of tail. **DISTRIBUTION** Found only in Central Highlands, in Pussellawa, Ramboda and Thalawakele areas. Endemic to Sri Lanka. **HABITAT AND HABITS** Found in secondary forests, open areas, roadsides and tea estates. Diurnal gecko that can be seen on boulders and rough walls, and under bridges. Females thought to be communal nesters.

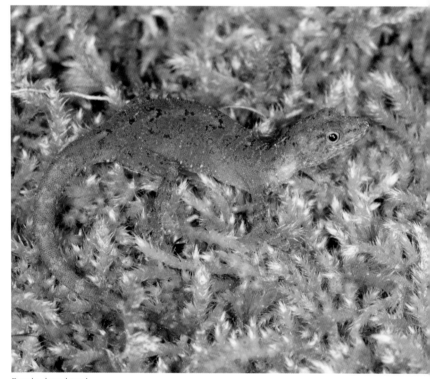

Female, dorso-lateral

Taylor's Forest Gecko ▪ *Cyrtodactylus edwardtaylori* 18cm Ⓔ
(*Sinhala* Taylorge Vakniyahuna)

DESCRIPTION Body large, stout and compressed. Head large and broad. Eyes large with vertical pupils. Limbs robust. Fingers slender, bent at an angle. Tail slender and long. Tubercles across mid-body, 14–15 rows. Dorsum and head light brown with only 4 dark brownish bands from back of head to base of tail. **DISTRIBUTION** Uva Province at 1,000m above sea level. Endemic to Sri Lanka. **HABITAT AND HABITS** Found in submontane forests, forest edges, tea estates and home gardens. Nocturnal and arboreal. Lives on tree trunks and in tree-holes 2–4m above the ground. Occasionally seen in houses.

Lighter

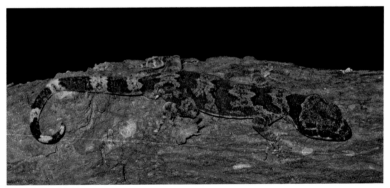

Darker

Great Forest Gecko ■ *Cyrtodactylus fraenatus* 20cm ⓔ
(*Sinhala* Mahakele Vakaniyahuna)

DESCRIPTION Body large, stout and compressed. Head large and broad. Eyes large with vertical pupils. Limbs robust. Fingers slender, bent at an angle. Tail slender and long. Enlarged paravertebral tubercle rows across body 17–23. Dorsum and head light brown, with more than 4 dark brownish bands (usually 5) from back of head to base of tail, which has about 8 dark bands. **DISTRIBUTION** Vicinity of Kandy, Gampola and Pussallawa areas at 500–800m above sea level. Endemic to Sri Lanka. **HABITAT AND HABITS** Inhabits forests, secondary forests, tea estates and home gardens. Nocturnal and arboreal. A slow moving species. Lives on tree trunks, in tree-holes and boulders, and in houses close to forests. Its tail is prehensile and is occasionally used to hold on to twigs. It is also moved like a wriggling worm to lure prey animals. Feeds on insects such as grasshoppers, bees, beetles and cockroaches, and tree frogs. Oviparous, laying about 2 hard-shelled, spherical eggs in crevices or soil. Adults can be seen basking for short durations in the mornings. Can bite savagely when caught.

Dorsal

Knuckles Forest Gecko ■ *Cyrtodactylus soba* 20cm ⓔ
(*Sinhala* DumbaraVakniyahuna)

DESCRIPTION Body large, stout and compressed. Head large and broad. Eyes large with vertical pupils. Limbs robust. Fingers slender, bent at an angle. Tail slender and long. Enlarged paravertebral tubercle rows on body 25–31. Posterior margins of ventral scales

pointed. Dorsum and head light brown to grey, with 5–6 dark brownish or greyish bands from back of head to base of tail, which has dark bands. **DISTRIBUTION** Restricted to Knuckles mountain range at 600–1,500m above sea level. Endemic to Sri Lanka. **HABITAT AND HABITS** Occurs in montane and submontane forests, pine plantations, tea estates, home gardens and insides of houses. Nocturnal and arboreal. Lives on tree trunks, in tree-holes and in houses close to forests. Feeds on crickets and cockroaches. Oviparous, laying 2 hard-shelled, spherical eggs in crevices, tree-holes or soil.

Lighter

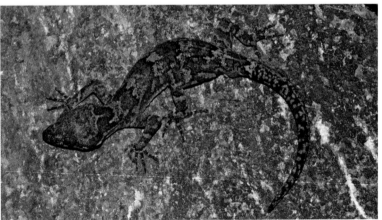

Darker

Spotted Bow-finger Gecko ■ *Cyrtodactylus trieda* 11cm **e**
(*Sinhala* Pulli Vakniyahuna)

DESCRIPTION Body medium sized and short. Head compressed and broad anteriorly. Fingers short and slender, bent at an angle and compressed at tips. Eyes large with vertical pupils. Tail shorter than head and body, and stumpy. Dorsum dark purplish-brown or grey, with or without yellow or white spots scattered on body from neck to tail-tip (no spots in regrown tail). **DISTRIBUTION**

Wet and intermediate zones, and wet regions in dry zone, at below 1,000m above sea level. Endemic to Sri Lanka. **HABITAT AND HABITS** Found in forests, forest edges and densely vegetated home gardens. Nocturnal, slow moving and terrestrial. Can be seen under stones and logs during day; at night, moving in leaf litter or at bases of tree trunks. Oviparous, laying 2 spherical eggs under stones or logs, or in loose soil.

Close-up of head

Dorso-lateral

Blotch Bow-finger Gecko ■ *Cyrtodactylus yakhuna* 8cm ⓔ
(*Sinhala* Yak Huna, Yak Vakaniyahuna)

DESCRIPTION Body medium sized and short. Head depressed and broad anteriorly. Fingers short, slender, bent at an angle and compressed at tips. Eyes large with vertical pupils. Tail shorter than head and body, and stumpy. Light brownish or cream with dark brown blotches on head and body; blotches usually have thin blackish border. Generally, 2 large, dark brown blotches or bands on trunk, and band on neck starts from behind eye on both sides (occipital band). Throat whitish with numerous blackish, elongated spots.

Darker form

Tail mostly lighter in colour, bearing scattered dark spots. **DISTRIBUTION** Semi-arid, dry and intermediate zones. Endemic to Sri Lanka. **HABITAT AND HABITS** Occurs in dry, mixed evergreen forests, scrubland and anthropogenic habitats. Nocturnal, slow-moving, ground dwelling species found among leaf litter and under logs. Curls tail in defence, exhibiting eye-like black spots. Oviparous, laying about 4 ovoid, hard-shelled eggs.

Lighter form with a curved tail displaying 'pseudo eyes'

Four-claw Gecko ■ *Gehyra mutilata* 11cm
(*Sinhala* Chathuranguli Huna)

DESCRIPTION Body medium sized, stout and depressed, with a lateral fold. Skin smooth. Tail carrot shaped and smooth, without any tubercles. Tips of digits broadly dilated and compressed, and with undivided lamellae. Claw in first inner digit minute. Dorsum light cream-brown to pale yellow with faint circular spots; colouration brighter in juveniles. **DISTRIBUTION** All over Sri Lanka, from sea level to 1,500m above. Extra-limital: widely distributed in many tropical countries in the world, from Madagascar to the Philippines; introduced to Australia, Mexico and the USA. **HABITAT AND HABITS** Found in houses, home gardens and plantations; very rarely seen in forests. Nocturnal gecko seen very commonly in houses. Easily seen near light bulbs preying on insects that come to lights in the night. Oviparous, laying 2 spherical, hard-shelled eggs that are fused together.

Dorsal

Kandyan Gecko ■ *Hemidactylus depressus* 14cm ⓔ
(*Sinhala* Hali Gehuna)

DESCRIPTION Body medium sized and depressed. Head large. Inner digit well developed; all digits with visible claws; tips of digits with divided lamellae. Pupils vertical. Skin rough. Paravertebral row has 38–43 tubercles; 13–16 longitudinal rows of mid-dorsal tubercles; no post-cloacal spurs; tail strongly depressed, with many spine-like tubercles. Dorsum light brownish to cream, with series of dark 'X' marks running from neck to base of tail.

Dark line bordered by lighter line on upper side runs across eye from tip of snout to angle of jaw. Distinct dark and light bands on tail. **DISTRIBUTION** Occurs all over Sri Lanka in all climatic zones, from sea level to more than 1,700m above. Endemic to Sri Lanka. **HABITAT AND HABITS** Found in forests and anthropogenic habitats in dry and intermediate zones, and houses and home gardens in wet zone. Nocturnal and arboreal. Can be seen on trees and in houses. Oviparous, laying 2 spherical eggs in crevice, among leaf litter or under bark.

Individual with darker markings

Individual with lighter markings

Common House-gecko ■ *Hemidactylus frenatus* 11cm
(*Sinhala* Sulaba Gehuna; *Tamil* Palli)

DESCRIPTION Body medium sized and depressed. Inner digit well developed; all digits with visible claws; tips of digits with divided lamellae. Pupils vertical. Dorsal surface of body smooth; tail rough, with spiny tubercles in anterior half; slightly depressed with round margin. Dorsum cream-brown to light brown, with scattered darker spots forming various patterns. Dark band runs across eye from snout to neck. **DISTRIBUTION** Throughout, in all climatic zones from sea level to 1,500m above. Extra-limital: widely distributed in many tropical and subtropical regions of the world, from Madagascar to Solomon Islands; introduced to Australia, New Caledonia, Fiji and tropical America. **HABITAT AND HABITS** Found in houses, home gardens and plantations; very rarely seen in forests of wet zone. Nocturnal and arboreal. Can be seen on trees and in houses. Easily observed near light bulbs preying on insects that are attracted to lights at night. Oviparous, laying 2 ovoid, hard-shelled eggs, which are attached to a substrate.

Light

Dark

Spotted Giant-gecko ■ *Hemidactylus hunae* 25cm ⓔ
(*Sinhala* Kimbul Huna/DavanthaTithhuna)

DESCRIPTION Body large, stout and depressed. Inner digit well developed; all digits with visible claws; tips of digits with divided lamellae. Pupils vertical. Skin rough; keeled dorsal tubercles arranged in 16–20 rows. Dorsum brownish-grey with 5 transverse bands with medially lighter light spots from head to base of tail, which is banded. Largest gecko in Sri Lanka. **DISTRIBUTION** Scattered locations of Eastern and Uva Provinces at 200–700m above sea level. Endemic to Sri Lanka. **HABITAT AND HABITS** Occurs in forests, grassland and home gardens adjacent to forests. Nocturnal and arboreal. Can be seen on large trees and boulders, and in caves and houses. Feeds on insects and smaller invertebrates. Bites savagely when captured. Oviparous, laying ovoid, hard-shelled eggs in leaf litter or crevices.

Typical darker colouration

Dorso-lateral

Termite-hill Gecko ■ *Hemidactylus lankae* 14cm ●
(*Sinhala* Humbas Huna)

DESCRIPTION Body large and depressed. Inner digit well developed; all digits with visible claws; tips of digits with divided lamellae. Pupils vertical. Large, knob-like tubercles from neck to base of tail, 16–20. Dorsum greyish-brown with dark cross-bars in which outer borders are lined with white, knob-like tubercles. Dark line bordered by 2 white dotted lines runs across eye from snout to back of head. **DISTRIBUTION** Semi-arid, dry and intermediate zones from sea level to 600m above. Endemic to Sri Lanka. **HABITAT AND HABITS** Found in forests, scrub jungle, open areas and home gardens. Nocturnal and terrestrial. Can be seen in termite mounds, boulders and rock piles, at the bases of trees and also in houses. Feeds on insects, spiders and worms. Oviparous, laying 2–6 ovoid, hard-shelled eggs in termite mounds and among leaf litter.

Close-up of head

Dorso-lateral

Bark Gecko ■ *Hemidactylus leschenaultii* 16cm
(*Sinhala* Kumbuk Huna)

DESCRIPTION Body large, stout and depressed. Inner digit well developed; all digits with visible claws; tips of digits with divided lamellae. Pupils vertical. Dorsal surface of skin without imbricate scales; dorsal tubercles irregularly arranged; skin smooth. Tail depressed, with spinous margin. Body colour highly variable: dorsum brownish-grey with large, dark brown patch running from neck to base of tail; some patches laterally connect, forming wavy line; tail barred. **DISTRIBUTION** Semi-arid, dry and intermediate zones from sea level to 300m above. Extra-limital: Pakistan and southern India. **HABITAT AND HABITS** Found in forests, woodland and home gardens. Nocturnal and arboreal. Can be seen on trees and boulders, and in caves and houses; noticeable basking on trees and buildings in the mornings. Feeds mostly on insects, but may also take other smaller geckoes and sizeable vertebrates. Oviparous, laying 2 ovoid, hard-shelled eggs.

Dorso-lateral

Dorsal

Spotted House-gecko ■ *Hemidactylus parvimaculatus* 11cm
(*Sinhala* Pulli Gehuna)

DESCRIPTION Body medium sized and depressed. Inner digit well developed; all digits with visible claws; tips of digits with divided lamellae. Pupils vertical. Dorsal surface of body without any keeled scales; skin rough with 16–20 rows of tubercles on body; tail rough, with scattered tubercles, and spiny tubercles on sides; 13–15 femoral pores on each side. Dorsum cream to light brown, with scattered small to medium-sized spots; tail may bear thin dark bands; dark stripe runs across eye from snout to back of

head. **DISTRIBUTION** Throughout, from sea level to 1,500m above. Extra-limital: South India. **HABITAT AND HABITS** Lives in houses, scrubland and home gardens. Nocturnal and arboreal. Commonly seen in houses. Also found under logs and stones. Commonly observed near lights in houses and on lamp-posts, preying on insects that are attracted to lights. Oviparous, laying 2 hard-shelled, ovoid eggs.

Dark

Light

Frill-tailed Gecko ■ *Hemidactylus platyurus* 11cm
(*Sinhala* Nagutavakarali Huna)

DESCRIPTION Body medium sized and depressed. All digits have visible claws; tips of digits with divided lamellae. Pupils vertical. Dorsal surface of body without keeled scales; tail highly depressed, with denticular margin; fingers with significant webbing. Dorsum light brown to dark brown with darker squarish marks in middle; darker lateral stripe from back of eye to groin; tail has faint cross-stripes. **DISTRIBUTION** Known from only two locations, in the lowland intermediate (Maspotha, Kurunegala) and dry (Kalmunai) zones. Extra-limital: India to Taiwan through Southeast Asia and China. **HABITAT AND HABITS** Inhabits old houses that are mainly built from laterite blocks, and associated anthropogenic habitats. Nocturnal and arboreal. Feeds on insects such as houseflies, mosquitoes, moths, crickets and cockroaches. Males are territorial and vocalize loudly to exclude other males. Oviparous, laying hard-shelled, spherical eggs; occasional communal oviposition also known.

Dorso-lateral

Indo-Pacific Slender Gecko ■ *Hemiphyllodactylus typus* 6cm
(*Sinhala* Sihin Huna)

DESCRIPTION Body small and slender. Tail long. Limbs comparatively short. Pupils vertical. Digit-tips moderately to broadly dilated; first inner digit small, clawless or with minute claw. Dorsum light brown with dark brown spots forming continuous line from head to base of tail on lateral sides of body. Yellow or orangish-yellow, 'W'-like mark on base of tail, which is generally lighter in colour; ventral aspect of tail yellowish or orangish-yellow. Some individuals may have a few small yellow or orangish-yellow spots scattered on body. **DISTRIBUTION** Wet and intermediate zones from sea level to 1,100m above. Extra-limital: Hawaii and French Polynesia, Pacific Rim islands, New Guinea, Indonesia, Malaysia and Indochina. **HABITAT AND HABITS** Occurs in forests, forest edges, tea estates, home gardens and insides of homes. Nocturnal and arboreal. Found on trees, under logs and stones, and in houses. Population in Sri Lanka comprises only females, hence considered to be a parthenogenetic population. Oviparous, with females laying 1–2 sperical eggs at a time.

Light

Dark

Scaly-finger Gecko or Mourning Gecko ■ *Lepidodactylus lugubris* 8cm
(*Sinhala* Salkapa Huna)

DESCRIPTION Body medium sized and depressed. Head large and broad. Pupils vertical. Tail broad and depressed. Digit tips moderately to broadly dilated; inner digit clawless.

Dorsum cream to light brown, with 'W'-shaped dark cross-bands running from neck to base of tail, which is banded. Dark line runs across eye. **DISTRIBUTION** Wet, intermediate and dry zones from sea level to 700m above. Extra-limital: Pacific Islands, Southeast Asia, South Asia, Australia, Thai-Malay peninsula, Mexico and other central American countries. **HABITAT AND HABITS** Inhabits forests, forest edges, home gardens and houses. Nocturnal and arboreal. Occurs on trees and in houses. During daytime, can be seen among leaf litter on forest floor. Some populations are parthenogenetic. Oviparous, laying 2–3 eggs.

Juvenile

Adult

Leschenault's Snake-eye Lizard ■ *Ophisops leschenaultii* 12cm
(*Sinhala* Panduru Katussa)

DESCRIPTION Body medium sized, cylindrical and slender. Head slightly pointed. Eyes large with transparent lower eyelids. Tympanum distinct. Limbs robust. Digits well developed. Tail slender and long. Scales: nostril between 2 nasal scales; 1 postnasal; prefrontals not separated by an interprefrontal; interparietal does not separate parietals. Dorsum has brown stripe from head to middle of of tail, and 2 thin black lines border the brown stripe; 2 whitish lines run parallel to black lines on sides; laterally, blank band runs from back of eye to base of tail, which is orangish-brown. Belly white. **DISTRIBUTION** Several populations scattered in lowlands of dry and intermediate zones, for example in plains of Uva Province, Udawalawa, Mullaitivu and Jaffna. Extra-limital: India. **HABITAT AND HABITS** Occurs in grassland and open woodland. Diurnal and fast-moving terrestrial lizard. Active in the mornings at 7–9.30 a.m. Feeds on insects such as flies and beetles. Can be seen among grasses and leaf litter, and basking and foraging on rock surfaces. Shuttles from shade to heated rock surfaces during day.

Dorso-lateral

SKINKS
The skinks are characterized by their smooth, shiny scales and large, symmetric, shield-like scales on the head. Their legs are relatively small, and some species have no limbs at all. They show tail autotomy and the tails can regenerate. More than 1,700 species are known from Africa, Europe, Asia, Australia, and North and South America. They were formally recognized in a single family, the Scincidae. However, recent studies place them in eight different families: Acontidae, Egerniidae, Eugongylidae, Lygosomidae, Mabuyidae, Ristellidae, Sphenomorphidae and Scincidae. Thirty-four species in eight genera are known from Sri Lanka. They are classified in the families Lygosomidae, Mabuyidae, Ristellidae, Sphenomorphidae and Scincidae. Twenty-six species are restricted to the island, and the genera *Chalcidoseps*, *Lankascincus* and *Nessia* are endemic to Sri Lanka.

Ocellated Skink ■ *Chalcides* cf. *ocellatus* 20cm
(*Sinhala* Thith Hikanala)

DESCRIPTION Body robust and elongated. Head pointed. Neck thick. Auditory canal opening visible. Limbs small, widely separated when adpressed. Tail long and tapering.

Close-up of head

Dorsal

Scales smooth and shiny. Nostril between rostral and nasal scale. Dorsum greyish-brown with series of white spots organized in black stripes running across body and tail. **DISTRIBUTION** A single specimen resembling *C. ocellatus* of southern Europe, the Middle East, Pakistan and India was discovered in a lowland rainforest in western Sri Lanka (Kalutara district). The specimen from Sri Lanka may belong to an entirely different species, but further studies are needed to confirm its taxonomic status. **HABITAT AND HABITS** Virtually nothing is known about the natural history of the Sri Lankan population. European and Middle Eastern populations are terrestrial and feed on arachnids, insects and small lizards. They inhabit coastal sand dunes, deserts and home gardens, and are ovoviviparous, giving birth to 4–10 young.

Four-toed Snake Skink ■ *Chalcidoseps thwaitesi* 15cm

(*Sinhala* Chathuranguli Sarpa Hikanala)

DESCRIPTION Body slender, long and cylindrical. Tail long and tapering. Head broad, pointed; tip blunt. Neck feebly distinct. Auditory canal opening visible. Limbs 4, very small, each with only 4 digits. Scales smooth and iridescent. Dorsum dark brown; belly light brown. **DISTRIBUTION**
Restricted to Knuckles
mountain range at 600–
1,000m above sea level.
HABITAT AND HABITS
Found in submontane and
riverine forests, cardamom
plantations and home gardens.
Semifossorial species occurring
under stones, decaying logs
and leaf litter. Diurnal, with
some crepuscular activity.
Feeds mainly on termites.
Oviparous, laying 2 soft-
shelled eggs in loose soil.

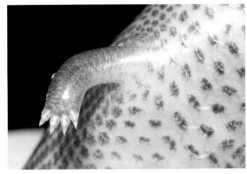

Close-up of forelimb

Dorso-lateral

Haly's Tree Skink ■ *Dasia haliana* 13cm ⓔ
(*Sinhala* Helige Ruk Heeraluwa)

DESCRIPTION Body robust and elongated. Head pointed. Neck distinct. Auditory canal opening visible. Limbs and digits elongated, digits 5. Tail long and tapering. Scales smooth and shiny; supranasal scales present; nostril in single nasal scale. Dorsum whitish-brown to white (in juveniles), with black transverse bands running across body from neck to tip

Dorsal

of tail. Head bears dark longitudinal stripes on top, as well as one running along eye. **DISTRIBUTION** Intermediate and dry zones of Sri Lanka from sea level to 700m above. Endemic to Sri Lanka. **HABITAT AND HABITS** Found in undisturbed and disturbed forests, and home gardens adjacent to forests. Diurnal and arboreal skink that also shows terrestrial habits. Feeds on insects and caterpillars. Oviparous, laying 2 soft-shelled eggs in loose soil.

Dorso-lateral

Common Skink ■ *Eutropis carinata* 30cm
(*Sinhala* Thamba Hikanala; *Tamil* Periya Arene)

DESCRIPTION Body robust. Head large and pointed. Neck indistinct. Auditory canal opening visible. Lower eyelids with scales. Limbs large, reaching wrist or elbow when adpressed. Scales not shiny; bear 3–5 keels; no postnasal scale. Dorsum bronze without any markings; light dorsolateral stripe runs along body starting behind eye; lateral sides orangish-red. Belly olive.

DISTRIBUTION All over Sri Lanka from sea level to 1,800m above. Extralimital: India, Bangladesh, Nepal and the Maldives. **HABITAT AND HABITS** Found in secondary forests, open areas, plantations and home gardens. Diurnal and terrestrial. Spends the night in crevices, and under stones and logs. Can be seen basking on boulders in the mornings. Feeds on insects, worms and centipedes. Oviparous, laying 2–20 soft-shelled eggs in loose soil.

Close-up of head

Dorso-lateral

Taylor's Striped Skink ▪ *Eutropis floweri* 15cm **e**
(*Sinhala* Taylorge Hikanala)

DESCRIPTION Body robust. Head pointed. Neck slightly distinct. Auditory canal opening visible. Lower eyelids scaly. Limbs well developed. Body scales with 3 keels; first pair of chin shields separated by single median scale; supranasals widely separated;

postnasal absent. Dorsum brownish-bronze with paired series of transverse black markings. Bright olive dorsolateral line runs from back of eye to base of tail. **DISTRIBUTION** Wet, intermediate and dry zones from sea level to 1,000m above. Endemic to Sri Lanka. **HABITAT AND HABITS** Inhabits forests and densely vegetated home gardens. Diurnal and terrestrial. Can be seen moving among leaf litter. Oviparous, laying 2 soft-shelled eggs under logs and stones.

Close-up of head

Dorso-lateral

Madarasz's Skink ■ *Eutropis madaraszi* 16cm ⓔ
(*Sinhala* Pulli Hikanala)

DESCRIPTION Body robust. Head pointed. Neck indistinct. Auditory canal opening visible. Lower eyelids scaly. Limbs well developed. Scales: dorsal scales with 3–7 keels; first pair of chin shields in medial contact; postsupralabial single; well-developed median keels under subdigital lamellae; paravertebral scales 34–42; prefrontals widely separated. Dorsum copper-brown with 6–8 longitudinal series of stripes; dorsolateral stripe one and a half scales wide starts from back of eye and ends midway in tail; lateral sides dark glossy brown. **DISTRIBUTION** Dry, semi-arid and intermediate zones from sea level to 200m above. Endemic to Sri Lanka. **HABITAT AND HABITS** Commonly seen in monsoon forests and adjacent home gardens. Diurnal species with semiarboreal tendencies. Frequently noted on rocks, boulders and tree buttresses, and moving among leaf litter.

Close-up of head

Dorso-lateral

Tammenna Skink ■ *Eutropis tammanna* 14cm ⊜
(*Sinhala* Thammanna Hikanala)

DESCRIPTION Body robust. Head short, broad posteriorly and pointed. Neck indistinct. Auditory canal opening visible. Lower eyelids scaly. Scales: dorsal scales with 3–7 keels; prefrontals in medial contact; first pair of chin shields in medial contact. Dorsum glossy chestnut, with scattered dark spots in posterior region; lateral sides dark glossy brown

Female, head

with scattered white spots; gular region dark orange in adult males. **DISTRIBUTION** Semi-arid, dry and intermediate zones from sea level to 300m above. Endemic to Sri Lanka. **HABITAT AND HABITS** Occurs in grassland, home gardens, coconut plantations and open woodland. Terrestrial and diurnal. Can be seen among leaf litter and on boulders. Feeds on insects. Oviparous, laying 2 soft-shelled eggs.

Male, dorso-lateral

Deignan's Lanka Skink ■ *Lankascincus deignani* 11cm ℮
(*Sinhala* Deignange Lakheeraluwa)

DESCRIPTION Body small and elongated. Tail long and cylindrical. Head pointed and broad anteriorly. Neck slightly distinct. Limbs small, all bearing 5 digits. Scales: ventral scales smooth; 2 frontoparietals; 2 loreals; primary temporal single; 7 supralabials; >50 paravertebral scales. Dark brown dorsally with 2 dark dorsolateral stripes that run from back of head to base of tail. In adult males, lateral sides and belly yellow, and throat black with white spots.

Lateral scales have white spots. **DISTRIBUTION** Wet and intermediate zones at 300–1,200m above sea level. Endemic to Sri Lanka. **HABITAT AND HABITS** Inhabits forests and densely vegetated home gardens. Diurnal and terrestrial. Can be seen among leaf litter, and under logs and stones. Can also bury itself in loose soil. Feeds on termites. Oviparous, laying 2 chalky-white, ellipsoid eggs in loose soil.

Female, dorso-lateral

Male, dorso-lateral

Catenated Lanka Skink ■ *Lankascincus dorsicatenatus* 12cm **e**
(*Sinhala* Damval Lakheeraluwa)

DESCRIPTION Body small and somewhat thick. Tail long and cylindrical. Head pointed and broad anteriorly. Neck slightly distinct. Limbs small, all bearing 5 digits. Scales: ventral scales smooth; 2 frontoparietals; 7 supralabials; last supralabial entire. Colour in adult males brown dorsally, with dark catenate stripe above each side and 1 laterally; throat region dark with each supralabial with white spot and 3 vertical rows of white spots behind ear. Belly yellowish and tail lighter in colour. **DISTRIBUTION** Wet zone at 30–1,000m above sea level. Endemic to Sri Lanka. **HABITAT AND HABITS** Occupies lowland rainforests, and densely vegetated home gardens and plantations. Diurnal, terrestrial and leaf litter-dwelling species. Regularly seen near water bodies.

Close-up of head

Dorso-lateral

Common Lanka Skink ■ *Lankascincus fallax* 11cm **e**
(*Sinhala* Sulabha Lakheeraluwa)

DESCRIPTION Body small and elongated. Tail long and cylindrical. Head pointed and broad anteriorly. Neck slightly distinct. Limbs small, all bearing 5 digits. Scales: ventral scales smooth; single frontoparietal. Dorsum light brown with lighter dorsolateral lines on sides. Two distinct colours for males: some have black throat with white spots and yellowish belly, others have red throat with white spots and yellowish belly. These colours may be due to changes in reproductive condition, though studies are needed to confirm this. **DISTRIBUTION** All climatic zones from sea level to 800m above. Endemic to Sri Lanka. **HABITAT AND HABITS** Found in open areas, monsoon forests, grassland, plantations (tea, coconut, rubber), and home gardens. Very common diurnal and terrestrial species. Usually seen among decaying leaf litter, rubble or grass, and also capable of burrowing in loose soil. Feeds on small insects, including termites. Oviparous, laying 1–2 eggs in loose soil or leaf litter, or under logs or boulders.

Male, black throat

Male, red throat

Female, white throat

Gans's Lanka Skink ■ *Lankascincus gansi* 11cm ⓔ
(*Sinhala* Gansge Lakheeraluwa)

DESCRIPTION Body small and thick. Tail long and cylindrical. Head pointed and broad anteriorly. Neck slightly distinct. Limbs small, all bearing 5 digits. Scales: ventrals smooth; frontoparietals 2; primary temporal double; supralabials 7; last supralabial split. Dorsum greyish-brown with brownish-black vertebral and flank stripes. Throat dark with

prominent white marks. Belly yellowish. Feet dark brownish. Red throat colour in males during breeding season. **DISTRIBUTION** Wet zone from sea level to 1,000m above. Endemic to Sri Lanka. **HABITAT AND HABITS** Found in forests and densely vegetated home gardens adjacent to forests. Diurnal and terrestrial. Can be seen among moist leaf litter, and under logs and stones. Oviparous, laying 1–2 eggs, which are buried in loose soil.

Adult male, dorso-lateral

Subadult, dorso-lateral

Munindradas's Lanka Skink ■ *Lankascincus munindradasai* 8cm ●
(*Sinhala* Munindradasage Lakheeraluwa)

DESCRIPTION Body small and elongated. Tail long and cylindrical. Head pointed and broad anteriorly. Neck slightly distinct. Limbs small, all bearing 5 digits. Scales:

frontoparietals 2; loreals 1; supralabials 6; primary temporals 2. Dorsum brownish with dark dorsolateral stripes starting from back of head and running to middle of tail. Females lighter in colour than males, which have bluish throat with blackish spots. **DISTRIBUTION** Known only from Sri Pada Sanctuary in Central Highlands at more than 1,000m above sea level. **HABITAT AND HABITS** Restricted to submontane and montane forests of Sri Pada Sanctuary. Terrestrial and diurnal. Can be seen among moist leaf litter, and under logs and stones.

Male, head

Adult male, dorso-lateral

Sripada Lanka Skink ■ *Lankascincus sripadensis* 12cm ●
(*Sinhala* Sripada Lakheeraluwa)

DESCRIPTION Body small and elongated. Tail long and cylindrical. Head pointed and broad anteriorly. Neck slightly distinct. Limbs small, all bearing 5 digits. Scales: 2 frontoparietals; primary temporal single; 7 supralabials; last supralabial entire; loreals 3; nasal divided. Copper-brown dorsally, with 2 blackish-brown dorsolateral stripes that run from back of head to beginning of tail. Lateral sides yellowish-brown. Tail brownish with dark spots.

DISTRIBUTION Known only from Sri Pada Sanctuary at 800–1,900m above sea level. Endemic to Sri Lanka. **HABITAT AND HABITS** Found in submontane and montane forests, and open areas. Diurnal, terrestrial and subfossorial. Can be seen among leaf litter and grass, and under logs and boulders. Adults can be seen basking in the mornings.

Close-up of head

Dorso-lateral

Smooth Lanka Skink ■ *Lankascincus taprobanensis* 12cm 🄴
(*Sinhala* Sumudu Lakheeraluwa)

DESCRIPTION Body small and elongated. Tail long and cylindrical. Head pointed and broad anteriorly. Neck slightly distinct. Limbs small, all bearing 5 digits. Scales: frontoparietals 2; supralabials 6; primary temporals 2; ventrals smooth. Dorsum dark brown; some individuals have dark spots that form longitudinal lines. Two dark dorsolateral lines run along body starting behind head and ending at tail. Lateral sides have white spots. Belly yellowish. **DISTRIBUTION**
Restricted to Central Highlands and Knuckles range at more than 1,000m above sea level. Endemic to Sri Lanka.
HABITAT AND HABITS Occurs in montane and submontane forests, montane grassland, home gardens, tea estates, and pine and eucalyptus plantations. Diurnal, terrestrial, semifossorial and leaf-litter dwelling. Can be seen among moist leaf litter, under logs and stones, and among grasses. Oviparous, laying 2 soft-shelled eggs among leaf litter or loose soil.

Close-up of head

Dorso-lateral

Taylor's Lanka Skink ■ *Lankascincus taylori* 9cm ⓔ
(*Sinhala* Taylorge Lakheeraluwa)

DESCRIPTION Body small and elongated. Tail long and cylindrical. Head pointed and broad anteriorly. Neck slightly distinct. Limbs small, all bearing 5 digits. Scales: 2 frontoparietals; 7 supralabials; single primary temporal scale; more than 50 paravertebral scales; 11–16 lamellae under fourth toe. Dorsum dark brown with dark dorsolateral stripe running from back of eye to base of tale. Males have dark throat with white spots. Belly yellow. **DISTRIBUTION** Wet and intermediate zones at 300–1,000m above sea level. Endemic to Sri Lanka. **HABITAT AND HABITS** Inhabits submontane and lowland rainforests, densely vegetated gardens and tea plantations. Diurnal and crepuscular, with terrestrial and subfossorial habits. Can be seen among leaf litter, and under logs and stones. Feeds on small insects such as termites. Oviparous, laying 1–2 eggs in loose soil.

Head

Dorso-lateral

Dotted Garden Skink ■ *Lygosoma punctatum* 17cm
(*Sinhala* Thith Heeraluwa; *Tamil* Arani)

DESCRIPTION Body small and highly elongated. Tail long, thick and cylindrical. Head pointed and broad anteriorly. Neck indistinct. Limbs small, all bearing 5 digits. Scales: 24–26 scale rows at mid-body. Dorsum creamish or sometimes bronze, with dark brown dots that form 4 longitudinal lines running from head to middle of tail; dots absent on sides of dorsum, forming thick, light-coloured line. Lateral sides heavily spotted. Juvenile colouration brighter, with tail bright crimson-red. **DISTRIBUTION** All climatic zones from sea level to 800m above. Extra-limital: India, Pakistan, Bangladesh and Nepal. **HABITAT AND HABITS** Found in scrub jungles, grassland, sandy areas, open woodland and home gardens. Common diurnal and subfossorial skink that can be seen among leaf litter, sand and loose soil, under logs and stones, and at bases of grasses. Fast-running skink that feeds on small insects. Oviparous, laying 2–4 eggs in loose soil.

Juvenile

Adult, dorso-lateral

Smith's Snake Skink ■ *Nessia bipes* 11cm **e**
(*Sinhala* Smithge Sarpaheeraluwa)

DESCRIPTION Body elongated, snake-like and cylindrical. Head pointed. No distinct neck. Eyes small. Tail short and cylindrical, with blunt-ended tip. Only 2 very short hindlimbs (as buds). Scales: very smooth and shiny; loreals 1 on each side; 28 scales at mid-body; interparietal scale broader than frontal. Dorsum reddish-brown or grey, with each scale with dark border. Belly lighter. **DISTRIBUTION** Knuckles mountain range at 500–1,000m above sea level. Endemic to Sri Lanka. **HABITAT AND HABITS** Occurs in forests, tea estates and home gardens. Fossorial, burrowing species that is nocturnal in behaviour. Found under logs and boulders, and among loose soil and leaf litter. Feeds on termites, earthworms and other fossorial invertebrates. Oviparous, laying 2–4 ellipsoid eggs in loose soil.

Dorso-lateral

Three-toe Snake Skink ■ *Nessia burtonii* 12cm ⓔ
(*Sinhala* Thriyanguli Sarpaheeraluwa)

DESCRIPTION Body elongated, snake-like and cylindrical. Head pointed. No distinct neck. Eyes small; lower eyelid movable with transparent disc. Tail long and cylindrical, with blunt-ended tip. Scales very smooth and shiny. Limbs: 4, very small and with 3 toes in each. Scales: 24 at mid-body; supranasals 3. Dorsum dark brown with darker edges on each scale. Belly lighter. **DISTRIBUTION** Wet zone from sea level to 800m above. Endemic to Sri Lanka. **HABITAT AND HABITS** Occurs in lowland rainforests, rubber estates and home gardens. Fossorial burrowing species that is nocturnal in behaviour. Found under logs and boulders and among loose soil and leaf litter. Feeds on termites and earthworms. Oviparous, laying 2 ellipsoid eggs in loose soil.

Close-up of limb

Dorso-lateral

Toeless Snake Skink ■ *Nessia monodactylus* 12cm **e**
(*Sinhala* Ananguli Sarpaheeraluwa)

DESCRIPTION Body elongated, cylindrical and snake-like. Head pointed. No distinct neck. Eyes small. Nostril placed in rostral scale. Limbs 4, without any digits. Scales very smooth and shiny. Tail short and cylindrical, with blunt-ended tip. Brown or grey with each dorsolateral scale bordered by darker margin. Belly pale. **DISTRIBUTION** Wet and intermediate zones at 300–800m above sea level. Endemic to Sri Lanka. **HABITAT AND HABITS** Occurs in forests and home gardens. Nocturnal and fossorial. Found under stones and decaying logs, and among leaf litter. Feeds on earthworms and termites. Oviparous, laying 1–2 ellipsoid, soft-shelled, pinkish-white eggs in loose soil. Hatchlings about 5cm in length.

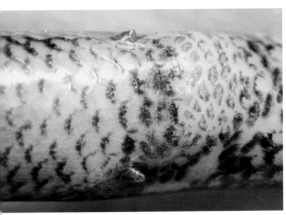

Limb buds

Dorso-lateral

Sarasin's Snake Skink ■ *Nessia sarasinorum* 13cm ⓔ
(*Sinhala* Sarasinge Sarpaheeraluwa)

DESCRIPTION Body elongated, cylindrical and snake-like. Head pointed. No distinct neck. Eyes small. Nostril placed in rostral scale. Two very short hindlimbs without any digits. Scales very smooth and shiny; loreals 2, mid-body 22. Tail short and cylindrical, with blunt-ended tip. Dorsum and tail brownish-grey. Belly lighter. **DISTRIBUTION** Intermediate and dry zones at 100–300m above sea level. Endemic to Sri Lanka. **HABITAT AND HABITS** Occurs in forests, open woodland and home gardens. Diurnal, fossorial and burrowing. Found in loose soil and leaf litter, and under logs and boulders. Feeds on insects such as termites and ants.

Ventral

Dorso-lateral

▪ Monitors ▪

Varanidae (Monitors)

Varanids are robust, diurnal lizards with long, non-autotomous tails and elongated necks. They have long, forked tongues that are used for chemoreception. Their sizes range from 20cm to 3m in length for the Komodo Dragon, the largest lizard in the world. They occupy a wide range of habitats, and can be surface dwelling, burrowing, arboreal or aquatic.

All monitors are fast hunters, and prey on small mammals, birds, birds' eggs, reptiles, amphibians and invertebrates. About 60 species in a single genus, *Varanus*, are known from Africa, central and southern mainland Asia, Malaysian and Indonesian islands, Papua New Guinea and Australia. Two species are known from Sri Lanka.

Land Monitor or Bengal Monitor ▪ *Varanus bengalensis* 150cm
(*Sinhala* Talagoya; *Tamil* Udumbu)

DESCRIPTION Large lizard with elongated snout; nostrils nearer to eye than snout-tip; long tongue. Body rather flat and bulky. Hindlimbs well built, and claws in both limbs sharp and long. Tail long and compressed, with serrated margin above. Juveniles yellowish-brown with yellow spots; adults dorsally dull brown or brownish-grey. **DISTRIBUTION** Widely distributed in all climatic zones up to 400m above sea level. Common in dry zone. Extra-limital: Afghanistan to Myanmar, India, Pakistan, Bangladesh and Nepal. **HABITAT AND HABITS** Found in forests, grassland, semi-deserts, plantations and human habitats. Diurnal, terrestrial and semiarboreal. Usually sleeps in tree-holes and house ceilings, generally emerging from hiding place during day, when temperature is high. Feeds on variety of insects, including beetles, spiders, snails, crabs, frogs, small mammals, birds, lizards and snakes, in addition to carrion. Oviparous, laying eggs in termitaria or tree-holes.

Adult

Juvenile

Water Monitor ▪ *Varanus salvator* 200cm
(*Sinhala* Kabaraya, Kabaragoya; *Tamil* Koyya)

DESCRIPTION Large lizard with elongated snout; long tongue that flickers out constantly. Body flat and bulky. Hindlimbs well built, and claws in both limbs sharp and long. Tail long and strongly compressed, with double-toothed, serrated margin above. Juveniles yellowish-brown with yellow spots; adults dorsally dull brown with dull yellow ocelli.
DISTRIBUTION Widely distributed in all climatic zones from coast to up to 500m above sea level. Extra-limital: Thailand, Indonesia, and Andaman and Nicobar Islands.

HABITAT AND HABITS Found in mangrove swamps, marshes, estuaries, rivers, lakes and canals in cities. Diurnal, terrestrial, aquatic and semiarboreal. Basks and swims during day. Feeds on insects, fish, crabs, turtle eggs, water birds and small mammals, in addition to carrion. Oviparous, laying 10–20 eggs inside termite mound or holes at bases of large trees. When cornered, distends neck, hisses and attacks with tail.

Juvenile

Adult

SNAKES
Sri Lanka is home to 104 snake species. This high diversity encompasses 11 families (Acrochordidae, Boidae, Colubridae, Cylindrophiidae, Elapidae, Gerrhopilidae, Homalopsidae, Pythonidae, Typhlopidae, Uropeltidae and Viperidae). Fifty are endemic, including two genera (*Aspidura* and *Balanophis*) Twenty-one are highly venomous and dangerous to humans; only six of these are terrestrial, while the rest are marine.

ACROCHORDIDAE (FILE SNAKES)
The Acrochordidae family consists of a single genus, *Acrochordus*, with three aquatic species. Two species occur in freshwater streams and estuaries in tropical East Asia and Australasia, while the other (*A. granulatus*) occurs in brackish and saline waters of tropical Asia and Australasia. File snakes possess small, tuberculate scales that outwardly terminate in small spines. This feature gives them a rough body surface, and also the name 'file snakes'. The tail is slightly compressed, and the body is laterally compressed when swimming. They are non-venomous and kill prey by constriction. Only one species occurs in Sri Lanka.

Little File Snake ■ *Acrochordus granulatus* 70cm
(*Sinhala* Diya-maha goda-maha, Mada Panuwa)

DESCRIPTION Body moderate in size. Head small, with short snout and small eyes. Nostrils on front upper surface of snout; neck not distinct. Body covered with small,

granular scales. Body slightly compressed and mid-body bulges, giving loose appearance to skin. Tail laterally compressed. Ventrals reduced. Body brown and grey with alternate pale buff bands. Sides have tinge of reddish-brown. In juveniles, bands are more distinct. Head contains a few cream or pale-coloured spots. Ventral aspect pale brownish-black. **DISTRIBUTION** Coastal Sri Lanka. Extra-limital: Pakistan to Fiji through India, South-east Asia and Australia. **HABITAT AND HABITS** Found in

Dorsal

lagoons, estuaries, mangroves and brackish water habitats along the coast. Nocturnal, Feeds mainly on bottom-dwelling fish. Ovoviviparous, producing 4–12 young. Non-venomous and inoffensive.

Ventral   *Close-up of head*

> ## CYLINDROPHIIDAE (PIPE SNAKES)
> Members of the Cylindrophiidae family have cylindrical bodies with indistinct head regions. They are mostly burrowing and subfossorial snakes. They have small heads, and small eyes with round pupils. Their tails are indistinct from the body and are extremely short and blunt. The family comprises two genera with 15 species distributed from Sri Lanka to Southeast Asia and southern China, excluding India.

Sri Lanka Pipe Snake ■ *Cylindrophis maculatus* 50cm ⓔ
(*Sinhala* Depath Naya, Wataulla)

DESCRIPTION Body medium sized and cylindrical. Tail very short and does not gradually taper. Neck not distinct. Scales smooth, shiny and iridescent in sunlight. Dorsal colour dull brick-red with black network enclosing 2 spots. Ventral side white with black bars or variegated pattern. **DISTRIBUTION** All over Sri Lanka in wet, intermediate and dry zones from sea level up to 600m above. Endemic to Sri Lanka. **HABITAT AND HABITS** Found in forests, plantations and home gardens. Nocturnal and subfossorial. During day hides under stones, decaying logs, rocks and leaf litter. Feeds on other smaller, non-venomous snakes, skinks and geckoes. Ovoviviparous, giving birth to 5–10 young. When disturbed, quickly flattens body, hides head, and raises and curls posterior quarter of body and tail. Non-venomous and inoffensive.

Dorso-lateral

Ventral

UROPELTIDAE (SHIELD-TAILED SNAKES)

The uropeltids are small, burrowing snakes. They have cone-shaped heads, often with a keel, and a cylindrical body with an indistinct neck region. Their tails are blunt, and some species bear a single large scale. They are non-venomous and ovoviviparous, giving birth to live young. Nearly 50 species are known, and they are restricted to Sri Lanka and southern India. Seventeen species in two genera (*Platypecturus* and *Rhinophis*) are known from Sri Lanka. All apart from one are endemic to Sri Lanka.

Blyth's Earth Snake ■ *Rhinophis blythii* 30cm ●
(*Sinhala* Gomara Thudulla)

DESCRIPTION Small snake with pointed head and snout. Body of uniform girth from head to tail. Neck not distinct. Tail short and ends in shield with minute blunt spines. Scales smooth and iridescent in sunlight. Ground colour of body brownish-black with yellow markings. Yellow 'V' mark encircles head, and tail region has distinct yellow area enclosing black patch dorsally. **DISTRIBUTION** Restricted to elevations more than 900m above sea level in Central Highlands of wet and intermediate zones. Endemic to Sri Lanka. **HABITAT AND HABITS** Nocturnal, fossorial and usually found under logs, stones, humus and decaying leaf litter, in forests, plantations and home gardens. Feeds mainly on earthworms. Ovoviviparous, giving birth to 3–6 young. Inoffensive, though like most uropeltids, when captured by hand it twists itself around the fingers and voids foul-smelling excreta.

Dorso-lateral

Drummond-Hay's Earth Snake ■ *Rhinophis drummondhayi* 30cm ⊖
(*Sinhala* Thapo Thudulla)

DESCRIPTION Small snake with pointed head and snout. Body of uniform girth from head to tail. Neck not distinct. Tail short and ends in shield with minute blunt spines. Scales smooth and iridescent in sunlight. Ground colour of body reddish-brown, with each scale with lighter margin; single row of scattered yellow scales on each side of body; tail has incomplete lighter ring with yellow markings. **DISTRIBUTION** Known only from elevations of 900–1,500m above sea level in southern part of Central and Uva Provinces (Haldumulla, Nanunukula). Endemic to Sri Lanka. **HABITAT AND HABITS** Usually found in decaying leaf litter in forests, silted drains in tea estates, humus piles near cattle sheds and grassland (Uva Pathanas). Nocturnal, fossorial and inoffensive. Feeds mainly on earthworms. Ovoviviparous, giving birth to 2–5 young.

Dorso-lateral

Eranga Viraj's Shield-tail Snake ■ *Rhinophis erangaviraji* 30cm ⊜
(*Sinhala* Eranga Virajge Thudulla; *Tamil* Eranga Virajvin nilakael Pambu)

DESCRIPTION Small snake with pointed head and snout. Body of uniform girth from head to tail. Neck not distinct. Tail short and ends in a shield with minute blunt spines. Scales smooth and iridescent in sunlight. Scale that makes shield of tail broader at end (caudal shield with one axis of symmetry) compared with that of Blyth's Earth Snake (see p. 94), the species it closely resembles. Ground colour of body dark brown or blackish, with yellow markings on dorsal and lateral sides. Dorsal and lateral surfaces of head are black. Belly has zigzag black pattern that runs along yellow background. No ring-like pattern at base of tail. Tail and anal region black. **DISTRIBUTION** Restricted to elevations above 900m in Rakawana mountain range in south-west wet zone. Endemic to Sri Lanka. **HABITAT AND HABITS** Fossorial burrowing species that frequents loose soil and leaf litter up to 1 m in depth in shady areas of tea estates, home gardens and grasslands, as well as natural forests of Rakwana mountain range. Non-venomous.

Dorso-lateral

Kelaart's Earth Snake ▪ *Rhinophis homolepis* 28cm ⓔ
(*Sinhala* Depath Thudulla)

DESCRIPTION Small snake with small head and pointed snout. Body cylindrical and of uniform girth from head to tail. Neck not distinct. Pointed rostral scale on head bears blunt keel. Tail short and ends in shield with minute blunt spines. Body blackish-brown to blackish-blue. Series of triangular yellow or orange spots runs along each side of body. Tail end is whitish pink and many bear two large black spots resembling two eyes (pseudo-eyes). **DISTRIBUTION** Wet zone at 400–800m above sea level. Endemic to Sri Lanka. **HABITAT AND HABITS** Common species found in plantations, home gardens and forests. Fossorial, and found in loose soil and leaf litter. Feeds on earthworms. Ovoviviparous, giving birth to 2–4 young. Inoffensive snake that excretes foul-smelling substance in defence. Non-venomous.

Tail end with two large 'eye-like' markings

Dorso-lateral

Gray's Earth Snake ■ *Rhinophis melanogaster* 25cm ⓔ
(*Sinhala* Kalu Wakatulla)

DESCRIPTION Body small and cylindrical. Head conical, bearing pointed rostral scale. Neck not distinct. Tail very short with spinose caudal shield ending in 2 points. Body usually blackish-brown with yellow patches and mottling on lateral side. Ventral

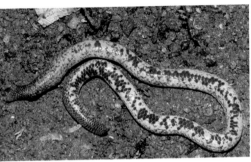

side black with yellow mottling. **DISTRIBUTION** Central Highlands 1,000m above sea level. Endemic to Sri Lanka. **HABITAT AND HABITS** Lives in loose soil and leaf litter, and under stones and decaying logs, in home gardens, plantations (banana, cardamom, *Pinus*, *Acacia*, tea), and submontane and montane forests. Nocturnal and fossorial. Feeds on earthworms and grubs. Ovoviviparous. Non-venomous, inoffensive snake that expels foul-smelling excreta as a form of defence.

Ventral

Dorso-lateral

Schneider's Earth Snake ■ *Rhinophis oxyrhynchus* 50cm ⓔ
(*Sinhala* Ulthudulla)

DESCRIPTION Small snake with small head and pointed snout. Body cylindrical and of uniform girth from head to tail. Neck not distinct. Pointed rostral scale on head bears distinct ridge. Tail short and ends in shield with minute blunt spines. Body dark to light brown, and belly region lighter, with ventral scales having yellow colour. **DISTRIBUTION** Lowland dry zone. Endemic to Sri Lanka. **HABITAT AND HABITS** Common species that can be found in home gardens, paddy fields, plantations, grassland and forests. Nocturnal and fossorial in habits. Feeds on earthworms. Very common in north Central Province, where it is killed during preparation of vegetable beds. Non-venomous.

Ventral

Dorso-lateral

Cuvier's Earth Snake ▪ *Rhinophis philippinus* 30cm ⓔ
(*Sinhala* Cuvierge Walga-abaya)

DESCRIPTION Small snake with pointed head and snout. Body cylindrical and of uniform girth from head to tail. Neck not distinct. Pointed rostral scale on head bears distinct ridge. Tail short and ends in shield with minute blunt spines. Scales smooth and iridescent in sunlight. Body uniformly bluish-grey to purplish-black or dark grey; head lighter. All dorsal and lateral body scales have thin white margins. **DISTRIBUTION** Wet and intermediate zones at 300–700m above sea level. Endemic to Sri Lanka. **HABITAT AND HABITS** Common species found in home gardens, plantations, grassland and forests. Nocturnal and fossorial. Feeds on earthworms and fossorial insects. Ovoviviparous, giving birth to 2–5 young. Non-venomous.

Dorsal

Phillips' Earth Snake ■ *Rhinophis phillipsi* 25cm ⓔ
(*Sinhala* Iri Wakatulla)

DESCRIPTION Body small and cylindrical. Head cone shaped, bearing keeled rostral scale. Neck indistinct. Tail very short with spinose caudal shield that ends in 2 points. Dorsum usually bluish-black with rows of yellow spots that run along dorsal and lateral sides. **DISTRIBUTION** Restricted to elevations more than 900m above sea level in Knuckles mountain range. Endemic to Sri Lanka. **HABITAT AND HABITS** Lives in loose soil in cardomum plantations, and submontane and montane forests, in Knuckles mountain range. Nocturnal and fossorial. Feeds on earthworms. Usually they are encountered at 10 to 15 cm below the surface in soil, and a few specimens can be encountered in a small area of 2 to 3 square metres Inoffensive and non-venomous. When picked up, they coil around fingers and defecate foul-smelling excrement.

Dorso-lateral

Large Shield-tailed Snake ▪ *Rhinophiss affragamus* 55cm ⓔ
(*Sinhala* Maha Bimulla; *Tamil* Mandalay Pambu)

DESCRIPTION Body small and cylindrical. Head pointed. Tail short, ending with distinctly flat, large scale that is covered with coarse spines. Body dark brown or blackish; ventral side lighter coloured, and iridescent in sunlight. Largest of the shield-tailed snakes in Sri Lanka.

Dorso-lateral

DISTRIBUTION Wet and intermediate climatic zones of southern lowlands. Endemic to Sri Lanka. **HABITAT AND HABITS** Lives in loose soil, humus heaps and under decaying logs. Nocturnal, fossorial, burrowing species that emerges during the night and forages on land. Feeds on earthworms. Ovoviviparous. Does not bite when handled. Non-venomous.

Dorso-lateral

▪ Pythons ▪

PYTHONIDAE (PYTHONS)

The Pythonidae family comprises medium-sized to large, muscular snakes that include the longest snake on Earth, the Reticulated Python *Malayopython reticulatus*, which can grow to more than 6m in length. They are non-venomous and kill their prey through constriction. Pythons are distributed from Africa through Asia to Australasia. They have heat-sensitive pits along the first parts of the upper lips (rostral and labial pits), which help in detecting prey. They also bear two spurs adjacent to the cloaca that are vestiges of the hindlimbs. About 40 species are known across their range, though only one occurs in Sri Lanka.

Indian Python or Rock Python ▪ *Python molurus* 300cm
(*Sinhala* Pimbura; *Tamil* Periya Pambu)

DESCRIPTION The largest snake in Sri Lanka. Body robust, cylindrical and ends in short tail. Head triangular. Distinct neck. Rostral and labial pits above lips. Dorsal body has irregular, dark golden-brown sub-quadrangular patches. Head has distinct lance-shaped mark. Dorsal scales smooth and iridescent in sunlight. Tail marbled in yellow and black. Belly white. **DISTRIBUTION** All over Sri Lanka except more than 1,200m above sea level. Extra-limital: Pakistan, India, Nepal, Bhutan, Bangladesh and Myanmar. **HABITAT AND HABITS** Occurs in primary and secondary forests, marshes, grassland and plantations. Nocturnal, terrestrial and arboreal snake that also show aquatic habits. Feeds on mammals, birds and monitor lizards. Oviparous, laying 10–50 eggs inside rock cave or large tree-hole. Some individuals are aggressive and may bite savagely when cornered. However, the species is non-venomous and kills prey through constriction.

Dorsal

Head with labial pits

Boidae (Boas)
Boas are medium to large snakes that include large examples like the anacondas of South America. They are characterized by their stout, muscular bodies. Most bear two spurs adjacent to the cloaca, which are vestiges of the hindlimbs. Roughly 60 species are distributed around Africa, Eurasia, Asia, New Guinea, Australia and the Americas. They are all non-venomous and kill prey by constriction. All are ovoviviparous, giving birth to live young. A single species occurs in Sri Lanka.

Rough-scaled Sand Boa ■ *Eryx conicus* 60cm
(*Sinhala* Kota Pimbura, Veli Pimbura; *Tamil* Mann Podeiyan)

Dorso-lateral

Ventral

DESCRIPTION Body short and thick; neck indistinct; tail short. Scales dull in appearance, and have prominent keels. Body colour a mixture of yellowish-brown with light grey mixed in. Large, dark brown, irregular blotches on dorsal aspect of body; lateral sides with irregular small blotches. Belly creamy-white. **DISTRIBUTION** Coastal dry and semi-arid zone of north-west, north and north-east to Yala. Extra-limital: Pakistan, India and Bangladesh. **HABITAT AND HABITS** Found in coastal sand dunes and scrubland. Nocturnal, terrestrial and sluggish. Feeds on lizards, rodents and birds, which it kills by constriction. Hides under sand, leaf litter, logs and stones during day. Ovoviviparous, giving birth to 3–8 young. If disturbed, usually hides head underneath body coils. Non-venomous, though some individuals may bite savagely.

COLUBRIDAE (TYPICAL SNAKES)
The so-called 'typical snakes' include a huge assemblage of snakes that vary in their mode of life, from burrowing to terrestrial, and arboreal to aquatic. There is therefore no set of uniform characteristics that define this group. Though the majority are harmless, a few have proven to have lethal bites, such as some keelback snakes; others, with their enlarged teeth, such as cat snakes and kukri snakes, need to be handled with care. More than 1,800 colubrid species in 320 genera are known worldwide. Sri Lanka is home to 42 species in 18 genera. Seventeen species are endemic to Sri Lanka, and the genera *Aspidura* and *Balanophis* are restricted to the island. Ten species in Sri Lanka have mildly toxic venom that is not harmful to humans, while one (*Balanophis ceylonensis*) has caused systemic envenoming.

Green Vine Snake ■ *Ahaetulla nasuta* 120cm
(*Sinhala* Ahaetulla; *Tamil* Kankuthi Pambu)

DESCRIPTION Long, slender snake with slightly compressed body. Tail cylindrical, tapering and long. Head elongated, and snout sharp and pointed. Neck distinct. Colour highly variable: common dorsal colour bright to dull green; ventral light yellow, green or dull purple/brown; some individuals have two light yellow or white lines along ventral side. **DISTRIBUTION** All over the island except at more than 1,000m above sea level. Extra-limital: India, Nepal, Bangladesh, Myanmar, Thailand, Cambodia and Vietnam. **HABITAT AND HABITS** Usually found on low vegetation, often near streams and other water sources. Diurnal and arboreal. Feeds on birds, frogs and lizards. Ovoviviparous, producing 4–20 young. When threatened, exhibits impressive threat posture by raising and inflating forebody, arching neck and opening mouth wide. Rear fanged and mildly venomous.

Typical green colouration

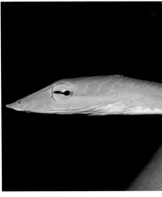
Close-up of head

Brown Vine Snake ■ *Ahaetulla pulverulenta* 100cm
(*Sinhala* Henakandaya)

DESCRIPTION Long and slender, with body slightly compressed. Tail cylindrical, tapering and long. Head elongated, and snout bears flat, scaly appendage. Distinct neck. General body colour light brownish-grey, mottled with dark brown spots; diffused dark lanceolate mark on head, and two dark brown lateral streaks across eyes. Ventral aspect light to dark chocolate-brown. **DISTRIBUTION** Throughout the island except at more than 1,000m above sea level. Extra-limital: India and Bangladesh. **HABITAT AND HABITS** Can be seen on low vegetation and hedges. Diurnal and arboreal. Feeds on geckoes, agamids and birds. Ovoviviparous. When threatened, raises and inflates forebody, arches neck and opens mouth wide. Mildly venomous.

Full body

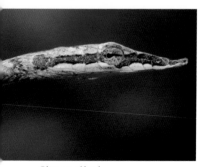

Close-up of head

Open mouth

Buff Stripe Keelback ■ *Amphiesma stolatum* 45cm

(*Sinhala* Aharakukka; *Tamil* Nikitan Kutti)

DESCRIPTION Body small and cylindrical. Tail cylindrical and tapering. Head distinct. Eyes large with round pupils. Scales keeled. Colour greyish-brown with 2 distinct buff stripes running along back from neck to tail-tip. Anterior half of body has regular dark brown cross-bars. Belly pearly-white. Two distinct colour variations have been noted.

DISTRIBUTION Widely distributed from sea level to 1,250m in all climatic zones of Sri Lanka. Extra-limital: Pakistan, India, Nepal, Bhutan, Bangladesh, Myanmar, Thailand Laos, Vietnam, China and Taiwan. **HABITAT AND HABITS** Frequents banks and bunds of ponds, streams, other waterways, paddy fields and open grass tracts. Diurnal and terrestrial. Feeds mainly on frogs. Oviparous, laying 5–15 eggs in crevices or termite mounds, or under decaying leaf litter. Docile and non-venomous.

Typical colouration

A mating pair, one with a red bands

Boie's Rough-side Snake ■ *Aspidura brachyorrhos* 40cm **e**
(*Sinhala* Le Madilla, Nidi Gulla)

DESCRIPTION Body small, slender and cylindrical. Tail short and tapering. Head slightly pointed, not distinct from neck. Eyes moderately sized with round pupils. Scales smooth, but males have spinous tubercles adjacent to cloaca (ischiadic region); costals 17; preocular present; subcaudals undivided; prefrontal does not touch eye. Colour pale yellow to blood-

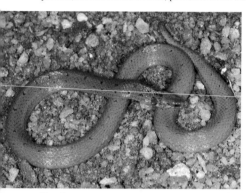

Dorsal

red above with 3–4 rows of longitudinal and vertebral rows of blackish dots. Nape bears oblique blackish spot on each side. Belly uniform yellowish to light orange; tail speckled with brown spots. **DISTRIBUTION** Wet and intermediate zones of Central and Uva Provinces at 300–900m above sea level. Endemic to Sri Lanka. **HABITAT AND HABITS** Found in forests, humid home gardens and tea estates. Nocturnal and semifossorial. During day found among decaying leaves and loose top soil, and under logs and stones. Feeds on earthworms. Oviparous, laying 2–5 eggs. Non-venomous and inoffensive.

Ventral

Black-spined Snake ■ *Aspidura ceylonensis* 40cm ⓔ
(*Sinhala* Rath Karawala, Kurun Karawala)

DESCRIPTION Body slender, long and cylindrical. Tail short. Head long, and snout broadly rounded. Neck slightly distinct. Scales smooth and iridescent; costals 17. Dorsal colour crimson-brown with black vertebral line. Dorsal side of forebody is brown. Lateral aspect has series of black spots in a line. Neck region has distinct dark brown marking. Chin and neck have tinge of yellow. Ventral aspect is crimson. **DISTRIBUTION** Wet zone of Sri Lanka at 600–1,500m above sea level. Endemic to Sri Lanka. **HABITAT AND HABITS** Lives in damp soil, in silted up drains, beneath heaps of decaying moist leaves, on tea estates, and under logs and stones in submontane forests. Subfossorial and nocturnal. Feeds mainly on earthworms. Oviparous, laying 3–5 eggs. Non-venomous and inoffensive.

Ventral

Dorso-lateral

Günther's Rough-side Snake ■ *Aspidura guentheri* 20cm ⓔ
(*Sinhala* Kuda Madilla)

DESCRIPTION Body small, slender and cylindrical. Tail short and tapering. Head slightly pointed, not distinct from neck. Eyes moderately sized with round pupils. Scales smooth, but males have feebly keeled scales in ischiadic region; costals 17, preocular present,

subcaudals undivided, prefontal touches eye. Colour light or dark brown dorsally with 3 longitudinal series of dark, light-edged dots. Head lighter above with yellowish collar around nuchal region. Belly light brown. **DISTRIBUTION** Elevations of 50–600m above sea level in wet zone of Sri Lanka. **HABITAT AND HABITS** Lives under stones and logs, in silt drains, in loose top soil and among moist leaf litter, in lowland rainforests, densely vegetated gardens and plantations. Semifossorial and nocturnal. Feeds on earthworms. Oviparous, laying 1–2 eggs in loose soil or under logs. Non-venomous and inoffensive.

Ventral

Dorsal

Common Rough-side Snake ■ *Aspidura trachyprocta* 30cm ⓔ
(*Sinhala* Dalawa Madilla)

DESCRIPTION Body small, slender and cylindrical. Tail short and tapering. Head slightly pointed, not distinct from neck. Eyes moderately sized with round pupils. Scales smooth and iridescent but males have feebly keeled scales in ischiadic region. Costals 15; preocular scale present; subcaudals undivided. Colour highly polymorphic; dorsal aspect may be uniform blackish, reddish-brown or in varying shades of yellowish-brown, with dark brown, red or black spots. Ventral aspect yellow or orange with dark spots. **DISTRIBUTION** Common, widely distributed form found in central hills more than 700m above sea level. Endemic to Sri Lanka. **HABITAT AND HABITS** Lives in moist decaying leaf litter and loose top soil, under stones and logs, in montane and submontane forests, tea estates, potato cultivation and home gardens. Nocturnal and subfossorial. Feeds mainly on earthworms and grubs. Oviparous, laying 5–10 eggs under decaying logs, rubble or leaf litter, or in loose soil. Non-venomous and inoffensive.

Red

Orange

Dark brown

Olive Keelback ■ *Atretium schistosum* 50cm
(*Sinhala* Diyawarna; *Tamil* Pachai Thanni-pambu)

DESCRIPTION Body medium sized and cylindrical. Neck barely distinct. Scales keeled; costals 19; ventrals 129–160, subcaudals 80–85. Dorsal colour olive-brown or green, usually with red-and-black lines along lateral sides of body. Belly has 3 distinct colour variations: cream, yellow and red. **DISTRIBUTION** Widely distributed at 30–700m above sea level in all climatic zones. Extra-limital: India and Nepal. **HABITAT AND HABITS** Occupies streams, ponds, wells and paddy fields. Diurnal, aquatic snake that is also quite at home on land. Feeds mainly on frogs and fish. Oviparous, laying 10–30 eggs inside crevices, under logs, in leaf litter and among stones near water sources. Non-venomous and inoffensive.

Ventral

Dorso-lateral

Sri Lanka Keelback ■ *Balanophis ceylonensis* 40cm ⓔ
(*Sinhala* Nihaluwa, Malkarawala)

DESCRIPTION Body cylindrical and moderately elongated. Head distinct from neck. Eyes large with round pupils. Lateral scales strongly keeled; costals 19; ventrals 137–143; subcaudals 40–54. Dorsal colour of adults dark brown with tinge of yellow, with diffused black spots running along lateral sides of body. Head darker with broad black stripe starting from eye. Juveniles have dark orange head and purplish-grey body, with parallel series of yellowish spots surrounded by black border that forms stripes along body. Ventral aspect whitish-cream. **DISTRIBUTION**

Confined to elevations of 30–900m above sea level in wet zone. Genus and species endemic to Sri Lanka. **HABITAT AND HABITS** Confined to lowland rainforests. Slow moving, diurnal and terrestrial snake that can be seen under logs and among leaf litter. Oviparous, laying about 4–5 eggs. Feeds on frogs. Generally docile but may bite occasionally. Mildly venomous, but a bite causing systemic envenoming is recorded.

Adult

Juvenile

Barne's Cat Snake ■ *Boiga barnesii* 60cm **ℯ**
(*Sinhala* Panduru Mapila)

DESCRIPTION Body medium sized, long and slender. Head distinct from neck. Large eyes with vertical pupils. Several different colour morphs. Common colour pattern is brown or yellowish-brown background, with broad purplish-brown bands or spots that run along body. Head lighter in colour but bears large, purplish-brown patch on top and lateral line

that runs from back of eye. Belly lighter, mottled with light brown spots. **DISTRIBUTION** Inhabits wet zone up to 1,000m above sea level. Endemic to Sri Lanka. **HABITAT AND HABITS** Occurs in lowland and submontane rainforests, secondary forests and occasionally home gardens. Arboreal. Feeds on geckoes and skinks. Oviparous, laying 4–5 eggs. Vibrates tail when disturbed. Generally not aggressive but may occasionally bite. Mildly venomous.

Dorso-lateral, typical colour

Grey morph

Beddome's Cat Snake ■ *Boiga beddomei* 90cm
(*Sinhala* Kaha Mapila)

DESCRIPTION Body medium sized, long and slender. Head distinct from neck. Large eyes with vertical pupils. Body generally yellow or orangish-yellow, with black or dark brown 'V' marks running along body. 'V' marks very faint in older specimens. Head bears faint lateral line starting behind eye, but lacks any other dark spots on dorsal aspect. Belly light yellow and slightly mottled.

DISTRIBUTION From sea level up to 700m above in dry, wet and intermediate zones. Extra-limital: Western Ghats of India. **HABITAT AND HABITS** Occurs in forests, secondary forests and occasionally home gardens; may enter houses at night. Nocturnal and arboreal. Feeds on geckoes and skinks. Oviparous, laying 5–10 eggs. Vibrates tail when disturbed. Mildly venomous.

Head, dorsal

Dorso-lateral

Sri Lanka Cat Snake ■ *Boiga ceylonensis* 100cm
(*Sinhala* Nidi Mapila)

DESCRIPTION Body medium sized, long, slender and slightly compressed. Tail long, cylindrical and tapering. Head large and subovate. Neck distinct. Eyes large with vertical pupils. Scales smooth and not shiny. Dorsal aspect greyish-brown with series of dark brown, irregular markings. Dark brown lateral stripe and large irregular brown spots on head.

Ventral aspect light brown with tinge of pink. **DISTRIBUTION** Common in all climatic zones up to 1,500m above sea level. Extra-limital: Western Ghats of India. **HABITAT AND HABITS** Frequently encountered in forests, secondary forests, plantations, home gardens and often inside houses. Nocturnal and arboreal. Feeds on geckoes, skinks, agamid lizards, birds and mice. Oviparous, laying 5–15 eggs. Vibrates tail, forms sinuous curves from forebody and lashes out to bite when disturbed.

Brown

Brown-orange

Forsten's Cat Snake ■ *Boiga forsteni* 140cm
(*Sinhala* Naga Mapila, Le Mapila)

DESCRIPTION Body large, long and well built. Tail long and tapering. Head subovate. Neck distinct. Eyes large with vertical pupils. Scales smooth and not shiny. Five distinct colour varieties. Most common variety dark grey with black-and-white cross-bars, and white belly. Most striking colour variety is uniform crimson-coloured form; with light crimson venter.
DISTRIBUTION Lowland dry, intermediate and wet zones. Extra-limital: India and Nepal.
HABITAT AND HABITS Occurs in forests, secondary forests and home gardens, and occasionally enters houses. Nocturnal and arboreal. Feeds on birds, agamid lizards, skinks and mice. Oviparous, laying about 5–15 eggs in tree-holes. Mildly venomous snake that usually rest inside tree-holes during daytime. When cornered, raises forebody slightly forming sinuous curves, hisses and vibrates tail.

Black and white

Crimson

Brownish

Gamma Cat Snake ■ *Boiga trigonata* 50cm
(*Sinhala* Raan Mapila; *Tamil* Poonai Pambu)

DESCRIPTION Body long, thin and compressed. Tail long and tapering. Head subovate and neck distinct. Eyes large with vertical pupils. Scales dull in appearance and not shiny. Dorsum dull yellow to greyish-brown, with series of light grey, black-edged, arrowhead-shaped markings. 'Y'-shaped mark on head; narrow dark streak bordered above with light grey runs from behind eye at an angle to mouth; dorsally there are 'V'-shaped light yellow

markings. Belly white or grey, speckled with dark grey spots. **DISTRIBUTION** Lowland wet and dry zones. Extra-limital: Iran, Afghanistan, Pakistan, India, Nepal and Bangladesh. **HABITAT AND HABITS** Inhabits forests, as well as tolerating disturbed environments such as parks and gardens, and also enters houses. Nocturnal and arboreal. Feeds on geckoes, agamid lizards and skinks. Oviparous, laying about 5–10 eggs. During day hides in tree-holes, stacks of bricks or wood. The most irritable Sri Lankan cat snake, attacking at the slightest movement. Mildly venomous.

Dorsal

Dorso-lateral

Ornate Flying Snake ■ *Chrysopelea ornata* 90cm
(*Sinhala* Malsara, Polmal Karawala; *Tamil* Parrakum Pambu)

DESCRIPTION Body slender and long. Tail cylindrical and long. Head depressed and pear shaped, and neck distinct. Eyes large. Scales rough and not shiny; some individuals have slight keels. Dorsal colour greenish-yellow with rosette marks between 2 black bars. Rosette marks most distinct from mid-body onwards. Ventral side pale greenish-yellow, and greenish-yellow costal scales have black border. Considered a colourful snake.

DISTRIBUTION Lowland wet zone. Extra-limital: India, Nepal, Bangladesh, Myanmar, Thailand, Malaysia, Cambodia, Laos, Vietnam, China and the Philippines.

HABITAT AND HABITS Found in lowland rainforests, plantations and occasionally home gardens. Diurnal and arboreal. Active snake that can leap from one branch to another, and glide down from trees by spreading its ribs and arching its body. Feeds on geckoes and other lizards. Oviparous, laying about 10 eggs. Some individuals are highly irritable and bite savagely, while others are docile. Mildly venomous, rear-fanged snake.

Head

Dorsal

Sri Lankan Flying Snake ■ *Chrysopelea taprobanica* 70cm
(*Sinhala* Dangara Danda)

DESCRIPTION Body medium sized, long and slender. Head depressed and oblong. Neck distinct. Eyes large with rounded pupils. Ventral scales with pronounced keels laterally; vertebral scales not enlarged; dorsal scales smooth or feebly keeled, with apical pits. Dorsum cream-brown with black zigzag bands outlined with thin yellow stripes. Head dorsally black with yellow stripes and spots. Belly cream or white. **DISTRIBUTION** Lowland intermediate and dry zone. Extra-limital: South-east India. **HABITAT AND HABITS** Found in forests, scrub jungles, cultivation and home gardens, and occasionally enters houses. Diurnal and arboreal. Has the ability to glide from one tree to another by flattening the body. Often descends to the ground. Feeds on lizards, especially geckoes and agamids. Oviparous, laying 5–10 eggs. Some individuals bite savagely. Mildly venomous.

Ventral

Dorsal

Trinket Snake ▪ *Coelognathus helena* 100cm
(*Sinhala* Katakaluwa, Mudu-habara; *Tamil* Kattu Pambu)

DESCRIPTION Body robust and long. Tail long, round and tapering. Head elongated and neck distinct. Large eyes. Scales smooth, not shiny. Dorsal aspect olive-brown; conspicuous white ocelli on lateral sides of forebody. On dorsal side 2 broad black lines run from behind head to mid-body. Black streak behind eye. Inside of mouth black. From mid-body on lateral aspects broad brown lines extend up to tail; ventral aspect pearly-white. **DISTRIBUTION** Widely distributed in all climatic zones up to 600m above sea level. Extra-limital: India, Pakistan, Nepal and Bangladesh. **HABITAT AND HABITS** Seen in forests and home gardens, and under decaying logs and rubble. Often enters houses at dusk. Diurnal and terrestrial, with occasional nocturnal habits. Feeds on rats and mice; juveniles feed on geckoes. Oviparous, laying 5–15 eggs. When agitated expands and raises forebody, forming sigmoid curves, opens black-coloured mouth and strikes. Non-venomous.

Dorsal

Dorso-lateral, defensive posture

Boulenger's Bronzeback ■ *Dendrelaphis bifrenalis* 60cm ⓔ
(*Sinhala* Panduru Haldanda, Haldanda; *Tamil* Komberi Mooken)

DESCRIPTION Body long, slender and cylindrical. Tail long, cylindrical and tapering. Head elongated, pear shaped and depressed. Neck distinct. Eyes large with round pupils. Scales are dull and not shiny. Dorsal colour uniform bronze with tinge of olive from head to mid-body; distinct black streak runs from snout to neck. Neck region has turquoise-blue tinge. Lateral sides have 2 cream-coloured lines with irregular black markings, and cream-coloured, paddy grain-shaped spots. Belly dull and light green. **DISTRIBUTION** Lowlands and mid-hills of wet, dry and intermediate zones. Endemic to Sri Lanka. **HABITAT AND**

Close-up of head

HABITS Lives in forests, scrubland and home gardens. Often encountered on low vegetation. Diurnal and arboreal, and may descend to the ground in search of food. When moving on the ground, has a peculiar habit of keeping head and an eighth of forebody erect. Feeds on frogs, geckoes, agamids and skinks. Oviparous, laying 4–12 eggs. Non-venomous, but may bite savagely when handled.

Dorso-lateral

Striped Bronzeback ■ *Dendrelaphis caudolineolatus* 75cm
(*Sinhala* Wairi Haldanda, Haldanda; *Tamil* Komberi Mooken)

DESCRIPTION Body long, slender and cylindrical. Tail long, cylindrical and tapering. Head elongated, pear shaped and depressed. Neck distinct. Eyes large with round pupils. Scales dull and do not shine. Dorsal colour uniform bronze with tinge of olive from head to mid-body; distinct black streak runs from back of eye to neck. Lower regions of head and neck area whitish. Dark, slightly thin, 'V'-shaped blackish cross-bars on anterior half of body. Belly lighter.
DISTRIBUTION Lowlands in wet and intermediate zones to 1,200m above sea level. **HABITAT AND HABITS** Found in lowland and submontane rainforests. Often encountered on low vegetation. Diurnal and arboreal. When moving on the ground, has peculiar habit of keeping head and eighth of forebody erect. Feeds on frogs, geckoes, agamids and skinks. Oviparous, laying about 3–6 eggs. Non-venomous.

Close-up of head

Dorso-lateral

Shokar's Bronzeback ■ *Dendrelaphis schokari* 100cm
(*Sinhala* Shokarige Haldanda, Haldanda)

DESCRIPTION Body long. Tail about a third of total length. Head distinct from neck. Eyes large with rounded pupils and golden irises. Scales smooth with apical pits. Dorsum unpatterned purplish- or bronzy-brown; neck and forebody yellow; buff flank-stripe from neck to vent; light blue on neck between scales that is revealed during display. Belly pale grey, green or yellow. **DISTRIBUTION** Lowlands and mid-hills of wet, dry and intermediate zones. Extra-limital: Western Ghats of south-western India. **HABITAT AND HABITS** Lives in open areas such as disturbed forests, forest clearings, and places in and around human habitations in rural areas. Diurnal and arboreal, but also known to forage on land. Can make long jumps between trees. Diet comprises frogs and lizards, as well as birds' eggs and insects. Oviparous, laying 6–7 eggs in tree-holes. Non-venomous.

Close-up of head

Lateral

Common Bronzeback or Seba's Bronzeback

■ *Dendrelaphis tristis* 80cm
(*Sinhala* Thuru Haldanda; *Tamil* Komberi Mooken)

DESCRIPTION Long and slender snake. Body elongate, slender and cylindrical. Tail long and tapering. Head elongate, pear shaped and depressed. Neck distinct. Eyes large with round pupils. Scales are dull. Head olive coloured with pale spot in middle. Forebody bronze with series of black streaks. Dorsal body colour olive; cream-spotted line along spine from neck to mid-forebody. Distinct black streak from behind eye to cheek. White line along ultimate costals at hind part of body. Belly dull yellowish-green. **DISTRIBUTION** Lowlands and mid-hills. Extra-limital: India, Pakistan, Bangladesh and Myanmar. **HABITAT AND HABITS** Lives in secondary forests, open areas, grassland, plantations and home gardens. Diurnal and arboreal. Feeds on frogs, geckoes, agamids and skinks. Oviparous, laying 6–12 eggs. Non-venomous species that may bite savagely.

Close-up of head showing the white spot

Lateral

Common Bridal Snake ▪ *Dryocalamus nympha* 40cm
(*Sinhala* Geta Radanakaya)

DESCRIPTION Body slender and cylindrical. Tail long, cylindrical and tapering. Head slightly elongated and subovate. Neck slightly distinct. Eyes large. Scales shiny and smooth, with 13 rows at mid-body. Dorsally black, or dark or light brown, with 50–60 white or cream cross-bars. Head bears large white band resembling bridal veil. Belly pearly-

Dorsal, brownish

white. **DISTRIBUTION** Lowland dry, semi-arid and intermediate zones. Extra-limital: India. **HABITAT AND HABITS** Occurs in forests, coconut plantations and rocky outcrops; occasionally enters human habitation in search of geckoes. Nocturnal and terrestrial (occasionally arboreal). Feeds on geckoes and skinks. Oviparous, laying about 3–5 elongated eggs. Knots body when handled. Non-venomous.

Dorsal, black

Reed Snake ■ *Liopeltis calamaria* 30cm
(*Sinhala* Punbariya)

DESCRIPTION Body slender. Tail long. Head elongated, with distinct neck. Pupils round. Dorsal colour a mixture of light to dark brown, with tinge of yellow. Belly creamy-white. Posterior portion of head has dark shading. Some individuals have 2 rows of pale black spots along spine (conspicuous anteriorly).
DISTRIBUTION Wet, dry and intermediate zones from sea level to 1,000m above. Extra-limital: South India. **HABITAT AND HABITS** Known to inhabit areas close to permanent water sources, forests and plantations. Uncommon diurnal and terrestrial snake. Active around dawn and dusk. Feeds on small frogs and geckoes. Oviparous. Non-venomous and inoffensive by nature.

Light

Darker

Common Wolf Snake ■ *Lycodon aulicus* 65cm

(*Sinhala* Alu Radanakaya, Alu polonga; *Tamil* Nai Pambu, Veedu Pambu)

DESCRIPTION Body cylindrical. Tail short, round and tapering. Head pear shaped, flat and distinct from neck. Scales smooth and shiny; 9 supralabials. Occurs in many colour forms. In common form, dorsal aspect of forebody is dark or light brown, merging with lighter shade posteriorly and series of about 20 white or cream cross-bars (which are prominent anteriorly) running along body; occasionally there are only 2–3 cream cross-bars. Head light brown. Belly pearly-white. **DISTRIBUTION** Wet, dry, semi-arid and intermediate zones from sea level to 1,000m above. Extra-limital: Pakistan, India, Nepal and Myanmar.

With 2-3 cross-bars

HABITAT AND HABITS Found in secondary forests and home gardens. Very common nocturnal and terrestrial snake. Hides under rubble, logs and leaf litter, inside crevices and underneath bark during day. Enters houses at night, so also called 'house snake'. Feeds voraciously on geckoes and skinks. Oviparous, laying 5–15 eggs. When disturbed, hides head underneath coils of body. Non-venomous but very aggressive species that bites savagely when handled.

Typical banded colouration

Sri Lanka Wolf Snake ■ *Lycodon carinatus* 50cm ⓔ
(*Sinhala* Dhara Radanayaka, Dhara Karawala)

DESCRIPTION Body small and cylindrical. Tail short and tapering. Head pear shaped, with round snout and distinct neck. Scales keeled and dull in appearance. Dorsal body black with distinct white rings, which may be thin or completely absent in mature individuals. Black colour extends to belly, but is fairly diffused. **DISTRIBUTION** Lowland wet zone from sea level to 700m above. Endemic to Sri Lanka. **HABITAT AND HABITS** Lives in lowland rainforests, plantations and home gardens. Nocturnal and terrestrial snake that is secretive by nature. Feeds on geckoes, skinks and other small, non-venomous snakes. Oviparous, laying 4–7 eggs. During day, hides under fallen leaves, logs and rubble. Moisture content in leaf litter in which it hides is a vital requirement for this snake. Mimics the Sri Lanka Krait (see p. 142). Hides head under body coils as a form of defence. Inoffensive and non-venomous.

Ventral

Dorso-lateral

Flowery Wolf Snake ■ *Lycodon osmanhilli* 40cm ⓔ
(*Sinhala* Mal Radanakaya)

DESCRIPTION Body small and cylindrical. Tail short, rounded and tapering. Head pear shaped and flat, and neck distinct. Scales smooth and shiny. Dorsally dark or light amber

with 15–20 light cream spots along vertebral line; spots occupy 3–4 scales and resemble small flowers. Labial scales cream, and belly pearly-white. **DISTRIBUTION** Lowlands of wet, dry, semi-arid and intermediate zones. Endemic to Sri Lanka. **HABITAT AND HABITS** Found in grassland, secondary forests and home gardens. Nocturnal and terrestrial. Feeds on geckoes and skinks. Females are oviparous. When disturbed, hides head underneath coils of body. Bites savagely when handled. Known to enter homes at night and killed on sight, as many believe it to be highly venomous, when it is in fact non-venomous.

Dorsal

Ventral

Shaw's Wolf Snake ■ *Lycodon striatus* 25cm
(*Sinhala* Kabara Radanakaya)

DESCRIPTION Body small and cylindrical. Tail short. Head small and neck not distinct. Scales smooth and shiny. Two colour variations occur: common variety dark brown with distinct light cream markings on dorsal aspect, which expand and diffuse laterally; in other individuals these white markings have slight yellowish-brown tinge. Belly pearly-white in both varieties. **DISTRIBUTION** All climatic zones from sea level to 1,000m above. Extra-limital: Iran, Pakistan, Afganistan, India and Nepal. **HABITAT AND HABITS** During day, hides under piles of rubble, bricks, logs and leaf litter, and at bases of grass clumps. Nocturnal and terrestrial. Feeds on geckoes and skinks. Oviparous, laying 2–5 eggs. When cornered, balls itself by throwing whole body into tight coils, and hides head underneath these coils. Inoffensive and non-venomous.

Dorsal

Common Kukri Snake ■ *Oligodon arnensis* 50cm
(*Sinhala* Arani Dathketiya; *Tamil* Yennai Panian, Olai Pambu)

DESCRIPTION Body slender and cylindrical. Tail short. Head ovate. Neck not distinct. Eyes have round pupils. Scales not shiny, but smooth. Dorsal body colour olive-green, greenish-grey or olive-brown, with 19–25 black cross-bars from neck to tail. Two chevron-shaped black cross-bars on head and neck; ventral aspect pearly-white. **DISTRIBUTION** Lowlands of wet, intermediate, dry and semi-arid zones up to 500m above sea level. Extra-limital: Pakistan, India and Nepal. **HABITAT AND HABITS** Occupies fringe areas of scrub and forests close to human habitats. Common diurnal and terrestrial snake with some crepuscular habits. Rests under heaps of rubble, decaying logs and leaf litter. Feeds on gecko eggs, young geckoes and skinks. Females lay eggs. Non-venomous snake that occasionally bites, inflicting an incised wound caused by its razor-sharp, kukri-like teeth.

Olive-green

Olive-brown

Templeton's Kukri Snake ■ *Oligodon calamarius* 25cm ⓔ
(*Sinhala* Kabara Dathkatiya)

DESCRIPTION Body small, cylindrical and slender. Tail short. Head ovate. Neck not distinct. Eyes have round pupils. Scales not shiny, but smooth. Dorsal body colour light brown to dark brown, with light vertebral stripe; 18–24 narrow dark brown, light-edged cross-bands that run either completely or halfway across back. Belly cream with square black spots. Head has dark crescent-shaped mark and elongated spot behind it. **DISTRIBUTION** Lowland wet zone from sea level up to 1,000m above. Endemic to Sri Lanka. **HABITAT AND HABITS** Occurs in lowland rainforests and secondary forests, and home gardens adjacent to forests. Diurnal and terrestrial. Feeds on gecko eggs. Females lay eggs. Coils into ball and hides head in body coils. Inoffensive and non-venomous.

Dorsal, dark brown

Ventral

Lateral, light brown

Dumerill's Kukri Snake ■ *Oligodon sublineatus* 25cm **e**
(*Sinhala* Pulli Dathketiya)

DESCRIPTION Body small, short and cylindrical. Tail short. Head ovate. Neck not distinct. Eyes have round pupils. Scales smooth. Dorsal side pinkish-brown with small, dark brown spots. Belly a mixture of light pink and brown, with 3 rows of brown spots; 2 lateral rows of linear marks confluent, forming stripes; median row of discontinuous spots ends at vent. **DISTRIBUTION** Found all over Sri Lanka up to 1,000m above sea level. Endemic to Sri Lanka. **HABITAT AND HABITS** Common terrestrial and diurnal snake. Usually hides under leaf litter, decaying logs and stones, and inside silted drains on estates. Pinkish-brown dorsal colour and dark brown spots help it merge with its habitat. Feeds mainly on reptile eggs and skinks. Oviparous, and female lays about 2 eggs. When threatened, twists itself into a ball. Non-venomous.

Dorso-lateral, dark

Ventral

Dorso-lateral, typical colouration

Variegated Kukri Snake ■ *Oligodon taeniolata* 38cm
(*Sinhala* Wairi Dathketiya)

DESCRIPTION Body small, short and cylindrical. Tail short. Head ovate. Neck not distinct. Eyes have round pupils. Scales smooth. Two subspecies occur, *O. t. ceylonicus* and *O. t. fasciatus*. Dorsal colour of *O. t. ceylonicus* is light reddish-brown or buff with light cream, broken vertebral line extending from neck to tail, black streaks radiating towards lateral sides, and dark brown mark on neck. Dorsum of *O. t. fasciatus* is light brown with narrow black transverse cross-bars that may form irregular spots. Dark band crosses snout from eye to eye in both subspecies. Ventral aspect pearly-white in both subspecies.

DISTRIBUTION Dry, semi-arid and intermediate zones from sea level to 500m above. *O. t. ceylonicus* is endemic to Sri Lanka, while *O. t. fasciatus* is also found in India and Pakistan. **HABITAT AND HABITS** Found in monsoon forests, scrub jungles, grassland and home gardens. Terrestrial and diurnal snake that is also active at dusk. Diet consists of lizard eggs and frog spawn. As a form of defence, arches neck, forming forebody into loose coils, and hides head underneath body. Non-venomous.

Oligodon taeniolata fasciatus

Oligodon taeniolata ceylonicus

Green Keelback ▪ *Macropisthodon plumbicolor* 50cm
(*Sinhala* Palabariya; *Tamil* Patchchai Pambu)

DESCRIPTION Body stout and cylindrical. Head subovate with round snout. Distinct neck. Tail short. Scales strongly keeled; costals 17–21; ventrals 142–161; subcaudals 27–45. Juveniles bright green dorsally; colour becomes dull as snake grows. Juveniles have distinct black markings on head and body that fade as they grow. Belly pearly-white.

Dorso-lateral, light colouration

DISTRIBUTION Wet, intermediate and dry zones from sea level to 800m above. Extra-limital: Pakistan, India and Myanmar. **HABITAT AND HABITS** Inhabits grassland close to water, open woodland, forests, tea plantations and home gardens. Nocturnal and terrestrial, with some semiarboreal and diurnal habits. During day hides under logs and stones. Feeds on frogs and toads. Oviparous, laying 10–20 eggs. When cornered, raises forebody and expands neck like a cobra; hence sometimes called 'green cobra'. Mildly venomous but inoffensive.

Dorso-lateral, dark colouration with few black spots

Indian Rat Snake ■ *Ptyas mucosa* 150cm
(*Sinhala* Gerandiya, Kahagerandiya, Kolagerandiya, Kalugerandiya; *Tamil* Sarai Pambu)

DESCRIPTION Body large, long, robust and cylindrical. Tail long, round and tapering. Head elongated and neck distinct. Eyes large with round pupils. Body scales smooth and shiny. Dorsal colouration highly variable: ranges from yellowish-brown and olivaceous-brown to black; posterior body has dark bands or reticulate pattern; belly greyish-white or yellow. Face lighter and bears black stripes that run along borders of scales.

DISTRIBUTION Throughout the island from sea level up to 2,000m above. Extra-limital: Iran, Afganistan, Pakistan, India, Nepal, Bangladesh, Myanmar, Thailand, Malaysia, Cambodia, Laos, Vietnam and China.

HABITAT AND HABITS Found in forests, agricultural fields, scrubland, mangroves and home gardens. Common, diurnal and terrestrial with strong arboreal habits. Feeds mostly on amphibians, rats, lizards and birds' eggs. Oviparous, laying 10–30 large eggs in termite mounds or tree-holes. When cornered, raises forebody and distends neck while hissing like a cobra, and bites savagely. Non-venomous.

Juvenile, lateral

Dorso-lateral, adult

Jerdon's Polyodont ■ *Sibynophis subpunctatus* 30cm
(*Sinhala* Dathi Gomaraya)

DESCRIPTION Body small and slender. Tail short. Head slightly distinct. Scales smooth. Dorsal side reddish-brown. Black spots along vertebral line. Dorsal colour of head dark brown to black; neck region cream with 2 black bands. Ventral aspect whitish with light green tint and dark spots. **DISTRIBUTION** All climatic zones from sea level to 500m above. Extralimital: India. **HABITAT AND HABITS** Found among leaf litter, near forest edges, scrubland, woodland and home gardens. Diurnal, terrestrial snake that also shows crepuscular activity. Feeds on geckoes, skinks and small, non-venomous snakes. Oviparous, laying 1–6 eggs.

Dorsal

Ventral

Sri Lanka Keelback or Boulenger's Keelback

■ *Xenochrophis asperrimus* 65cm **e**
(*Sinhala* Diya Bariya, Diya Polonga)

DESCRIPTION Body medium sized, robust and cylindrical. Tail one-third of body length. Head slightly oval. Neck distinct. Scales strongly keeled; costals 19–21; ventrals 131–146; subcaudals 73–93. Anterior half of body pale olive or reddish, with 2 series of distinct round or rhomboidal dark spots; posterior half may be dark olive or brownish with small blackish spots. **DISTRBUTION** Dry, wet and intermediate zones from sea level to 1,000 above. Endemic to Sri Lanka. **HABITAT AND HABITS** Inhabits waterways, streams, rivers, ponds, lakes, paddy fields and even domestic wells. Hides in crab holes or crevices in stream or river banks. Aquatic and diurnal with some nocturnal habits. Feeds mainly on frogs and fish. Oviparous, laying 10–30 eggs. Non-venomous species that is irritable and bites without hesitation.

Close-up of anterior region

Dorso-lateral

Checkered Keelback ■ *Xenochrophis* cf. *piscator* 75cm
(*Sinhala* Diya Naya; *Tamil* Thanni Pambu)

DESCRIPTION Body medium sized, robust and cylindrical. Tail one-third of body length. Head slightly oval, with distinct neck. Scales strongly keeled; costals 19–21; ventrals 126–143; subcaudals 72–94. Colouration highly variable: common dorsal colour olive-amber with series of faint checks that are more conspicuous anteriorly; 2 black streaks run obliquely directed downwards below eye. Belly pearly-white. **DISTRIBUTION** Widely distributed from lowland plains up to elevations above 1,200m. Extra-limital:

Afghanistan, Pakistan, India, Bhutan, Nepal, Bhutan, Bangladesh, Myanmar, Thailand, Malaysia, Vietnam, China and Taiwan. **HABITAT AND HABITS** Frequents waterways, streams, rivers, ponds, lakes, paddy fields and even domestic wells. Semiaquatic, spending most of its life in water. Predominantly diurnal with some nocturnal activity. Feeds on frogs and fish. Oviparous, laying 10–45 eggs in hole or crevice in embankment. When cornered, flattens neck region and forepart of body, and may bite savagely. Non-venomous.

Dorsal aspect with faint checks

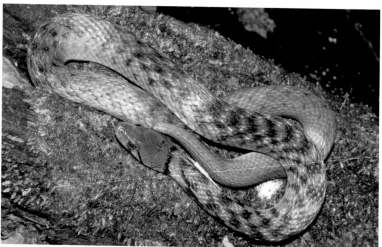

With checks

> **ELAPIDAE (COBRAS, CORAL SNAKES, KRAITS, MAMBAS AND SEA SNAKES)**
> Snakes of the Elapidae family are characterized by the presence of fixed, short, hollow
> fangs in the front of their mouths. They have slender bodies with shiny scales, and
> round pupils. Elapids are venomous and some bear extremely toxic venom. Their
> size ranges from 18cm in crowned snakes, to 4–5m in large King Cobras. About 270
> species are known worldwide. The family includes terrestrial, burrowing, arboreal and
> aquatic species. Most terrestrial species are oviparous and some are ovoviviparous. Two
> groups have independently colonized the sea, giving rise to oviparous 'sea kraits' and
> ovoviviparous 'true sea snakes'. In Sri Lanka there are five species of terrestrial elapid
> (in three genera), and 15 species of true sea snake (in two genera).

Common Krait ■ *Bungarus caeruleus* 90cm

(*Sinhala* Thel Karawala, Magamaruwa; *Tamil* Yettadi Viriyan, Karuwelen Pambu)

DESCRIPTION Body medium sized, elongate and cylindrical. Head indistinct from neck and bears small eyes. Scales are shiny; vertebral scales enlarged and hexagonal in shape. Dorsal body steely-blue, black or dark brown, with paired narrow white bands across it; in old individuals these are reduced to vertebral spots or disappear completely. Belly non-patterned and white. **DISTRIBUTION** Lowland semi-arid, dry and intermediate zones. Extra-limital: Afghanistan, Pakistan, Nepal, India and Bangladesh. **HABITAT AND HABITS** Found in thinly wooded forests, grassland, agricultural fields and human habitations. Nocturnal and terrestrial. During day hides under piles of debris, leaf litter, stones and logs. Feeds on mice, frogs and lizards, but prefers other small snakes. Oviparous, laying 6–15 eggs. Highly venomous. Not aggressive during day, though at night may bite without provocation.

Dorsal with paired white bands

Adult without white bands

Sri Lanka Krait ■ *Bungarus ceylonicus* 75cm ⓔ
(*Sinhala* Mudu Karawala)

DESCRIPTION Body medium sized and cylindrical. Head indistinct from neck. Eyes small and black. Body scales shiny, and vertebral scales enlarged and hexagonal in shape. Dorsal body steely-blue or black, with thick white or yellow bands along it. Belly has same dorsal pattern but in lighter intensity. **DISTRIBUTION** Wet and intermediate zones from sea level to about 1,800m above. Endemic to Sri Lanka. **HABITAT AND HABITS** Inhabits primary and secondary forests, plantations and home gardens. Nocturnal and terrestrial. During day hides among leaf litter, and under stones and logs. Feeds on smaller snakes, skinks and young mice. Oviparous, laying 5–10 eggs. Not an aggressive species, but highly venomous.

Faint bands

Ventral

With bands

Slender Coral Snake ■ *Calliophis melanurus* 30cm
(*Sinhala* Depath Kaluwa; *Tamil* Pattai Kattu Viriyan)

DESCRIPTION Body small, elongated, cylindrical, and equal in diameter from head to tail. Tail short. Neck indistinct. Dorsal body colour sandy-brown, and every scale has tiny brown dot. Head and neck black with 2 cream or white spots on back of head. Tail has 2 black rings. Belly yellowish or orangish, and tail light blue. **DISTRIBUTION** Lowland dry, intermediate and semi-arid zones. Extra-limital: India and Bangladesh. **HABITAT AND HABITS** Terrestrial snake that is active at dusk. During day it hides under leaf litter, decaying logs and stones. Feeds on small snakes. When disturbed coils body, and often hides head underneath. Also curls tail over back and exposes red-and-blue patches on ventral aspect of body. Moderately venomous, though its bite has not caused any deaths or serious problems in Sri Lanka.

Dorsal

Displaying the brightly coloured tail

Spectacled Cobra or Indian Cobra ■ *Naja naja* 150cm
(*Sinhala* Naya, Nagaya; *Tamil* Nalla Pambu)

DESCRIPTION The only snake with a distinct hood in Sri Lanka. Body large. Hood has spectacle-like marking on dorsal side, though this mark is highly variable and some individuals do not have markings; however, 2 black spots on ventral aspect of hood are constant. Head distinct from body, which is greyish-brown or brownish-black with thin, irregular white stripes. Ventral side light coloured with light grey bands, which are distinct on first half of body. **DISTRIBUTION** All over Sri Lanka except elevations more than 1,200m above sea level. Extra-limital: Pakistan, India, Nepal, Bhutan, Bangladesh and Myanmar. **HABITAT AND HABITS** Found in plantations, forests, grassland, paddy fields and also home gardens. Generally crepuscular, but may also be diurnal or nocturnal. Feeds on small mammals, frogs, birds, lizards, birds' eggs and other snakes. Oviparous, laying 12–30 eggs in April–July; female guards nest until eggs hatch about 60 days later. Expands hood and hisses loudly when disturbed. Highly venomous.

Dorsal hood

Ventral hood

Shaw's Sea Snake ■ *Hydrophis curtus* 65cm
(*Sinhala* Shawge Kuda Muhudu-naya)

DESCRIPTION Body stout and short. Head relatively large, broad and short. Body scales square or hexagonal; lowermost scale row with short keel. Dorsally brownish-grey to olive, with lighter bands that taper on sides. Belly yellow, white or cream. Females larger and heavier than males and in females, lowermost scale row is smooth.
DISTRIBUTION Common in north, north-west and east. Extra-limital: Persian Gulf to Australasia, through South and Southeast Asia and southern China. **HABITAT AND HABITS** Inhabits seas with muddy bottoms. Diet very broad, consisting of fish, crustaceans and gastropods. Ovoviviparous, giving birth to 1–6 young in May–September. Commonly caught in fishing nets in Sri Lanka. Highly venomous.

Spines on ventral side

Dorso-lateral

Annulated Sea Snake ■ *Hydrophis cyanocinctus* 120cm
(*Sinhala* Wairan Muhudu-naya; *Tamil* Kadal Pambu)

DESCRIPTION Long, elongated snake. Tail flattened and paddle-like. Head indistinct from neck; mid-body is thickest part of body. Ventral scales slightly broader than body scales. Ground colour typically olive or yellow, with dark transverse bands that may or may not encircle body. Lighter interspaces between bands thinner than darker bands.

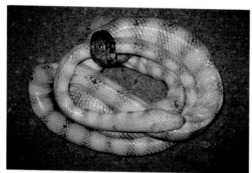

Belly yellowish-cream or pale yellow, with keeled scales. **DISTRIBUTION** Coastal waters throughout Sri Lanka. Extra-limital: coasts of Persian Gulf, South Asia and Southeast Asia. **HABITAT AND HABITS** Common species that inhabits shallow coastal seas. Diet includes fish and especially eels. Ovoviviparous, giving birth to 3–18 young. Regularly caught in fishing nets. Highly venomous.

Ventral

Dorso-lateral

Striped Sea Snake ■ *Hydrophis fasciatus* 100cm

(*Sinhala* Kudahis Muhudu-naya; *Tamil* Kadal Pambu)

DESCRIPTION Medium-sized, elongated snake with small head. Tail flattened and paddle-like. Ventral scales slightly broader than body scales, and thickest part of body is mid-body region. Head and neck region shiny black with oval yellowish spots. Rest of body shiny black to dark olive, with large, rhomboidal grey or blackish spots that may extend down sides, forming rings. **DISTRIBUTION** Only seen in coastal regions of northern peninsula of Sri Lanka. Extra-limital: coasts of India, Bangladesh, and Myanmar to Straits of Malacca. **HABITAT AND HABITS** Uncommon species that inhabits shallow seas about 5km from shore. Feeds on eels. Ovoviviparous, giving birth to 4–7 young. Occasionally caught in fishing nets in Sri Lanka. Highly venomous.

Ventral

Dorsal

Jerdon's Sea Snake ■ *Hydrophis jerdoni* 100cm
(*Sinhala* Jerdonge Muhudu-naya; *Tamil* Kadal Pambu)

DESCRIPTION Body medium sized and robust. Tail flattened and paddle-like. Head short, and not distinct from rest of body, but it is pointed when viewed dorsally. Ventral scales slightly broader than body scales. Body scales strongly keeled. Supralabials 6. Head and body may be yellowish, olive or light greenish, with black dorsal spots that extend around body to form complete bands. Tail has same body colour, but tip is black. **DISTRIBUTION** Northern, north-western and eastern coasts of Sri Lanka. Extra-limital: coastal waters of South and Southeast Asia. **HABITAT AND HABITS** Occurs in shallow seas about 4–10km from shore. Uncommon species in Sri Lanka. Feeds on snake eels. Occasionally caught in fishing nets. Highly venomous.

Head

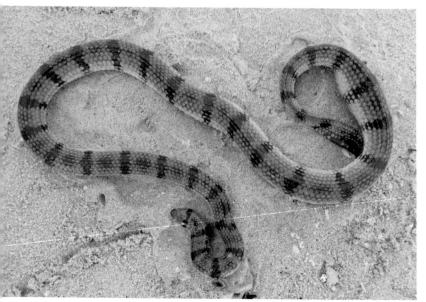

Dorsal

Arabian Gulf Sea Snake ■ *Hydrophis lapemoides* 90cm
(*Sinhala* Persiyanu-bokke Muhudu-naya)

DESCRIPTION Body medium sized. Tail flattened and paddle-like. Head somewhat distinct from neck. Ventral scales slightly broader than body scales; keels of ventral scales spike-like. Head darker in colour than body, with angular lighter region with apex at snout. Body colour creamish-yellow or pale grey, which further fades towards belly region;

32–35 dark grey bands on body that are usually broad dorsally, narrower and lighter laterally. **DISTRIBUTION** Northern and north-western coasts of Sri Lanka. Extra-limital: coastal regions of Arabian Gulf to western coasts of Thailand. **HABITAT AND HABITS** Inhabits shallow coastal waters with seagrass beds. Active at night and in early morning. Feeds on fish. Ovoviviparous, giving birth to about five young. Generally inoffensive, but becomes irritable when taken out of water. Highly venomous.

Dorso-lateral

Dorsal aspect of the rough-scaled form

Ornate Sea Snake ■ *Hydrophis ornatus* 100cm
(*Sinhala* Grayge Muhu-dunaya Muhudu-naya; *Tamil* Kadal Pambu)

Dorso-lateral

DESCRIPTION Body medium sized. Tail flattened and paddle-like. Head distinct from body, and thickest part of body is twice size of neck. Ventral scales slightly broader than body scales. Head and body greyish, pale olive or white, with dark grey bars or spots that are interspaced by thin lines. Ventral side cream in colour. **DISTRIBUTION** Northern and south-eastern coastal waters. Extra-limital: coastal regions from Arabian Gulf to Southeast Asia through South Asia. **HABITAT AND HABITS** Feeds on fish. Occasionally caught in fishing nets that are laid 4–5km off coast. Highly venomous.

Dorsal

Beak Sea Snake or Hook-nose Sea Snake

■ *Hydrophis schistosa* 100cm
(*Sinhala* Valakaddiya; *Tamil* Valakadiyan)

DESCRIPTION Large in size. Tail flattened and paddle-like. Head large; mid-body thickest part of body. Ventral scales slightly broader than body scales. Rostral scale at tip of snout resembles small beak. Head, body and tail dorsolaterally greyish-green or olive-brown, with belly region being lighter in colour. Some individuals have slightly darker bands on body, which are pronounced in juveniles.
DISTRIBUTION Brackish water habitats off the coast. Extra-limital: coastal regions of South Asia, Gulf of Arabia and Southeast Asia. **HABITAT AND HABITS** Brackish water species that occurs in coastal lagoons, estuaries and bays. Feeds on fish (mainly catfish). Ovoviviparous, giving birth to 7–16 young. Commonly caught in fishing nets, and becomes aggressive and bites when handled. Highly venomous.

Close-up of head

Dorso-lateral

Yellow Sea Snake ■ *Hydrophis spiralis* 170cm
(*Sinhala* Maha Muhudu-naya; *Tamil* Kadal Pambu)

DESCRIPTION Body large and elongated. Tail flattened and paddle-like. Head and neck slender compared with rest of body, which becomes thicker towards posterior. Body scales feebly imbricate or smooth. Body and head dorsally olive-yellow to olive-brown,

with 35–50 encircling dark bands that are narrower than their interspaces. Black dorsal spot between bands may be present. Belly yellow, and tail yellowish with dark bands and dark tip. **DISTRIBUTION** North, north-west, east and south-west of Sri Lanka. Extralimital: coasts of Arabian Gulf and South Asia to Straits of Malacca. **HABITAT AND HABITS** Occurs in shallow seas about 4–10km from shore, commonly caught in fishing nets in Sri Lanka. Feeds mainly on eels. Ovoviviparous, giving birth to 5–14 young. Very aggressive when handled or caught in fishing nets. Highly venomous.

Ventral

Dorsal

Yellow-belly Sea Snake ■ *Hydrophis platurus* 70cm
(*Sinhala* Bada-Kaha Muhudu-naya; *Tamil* Kadal Pambu)

DESCRIPTION Body short, medium sized and greatly compressed. Tail flattened and paddle-like. Head elongate and flattened. Neck distinct. Ventral scales irregular in shape and indistinct from rest of body scales. Dorsal body colour variable; common colours are black or dark brown. Belly light brown or yellow. Tail usually yellow or white with black spots. **DISTRIBUTION** Pelagic waters right around Sri Lanka. Extra-limital: the most widely distributed snake species in the world.

Occurs in seas of Indian and Pacific Oceans, including in certain temperate regions. **HABITAT AND HABITS** This pelagic species occurs commonly in oceanic drift lines, but is rarely caught in fishing nets in Sri Lanka. Feeds on fish that occur around floating vegetation on surface or close to it. Ovoviviparous, giving birth to young. Aggressive and highly venomous.

Close-up of head

Dorso-lateral

Viperine Sea Snake ■ *Hydrophis viperinus* 80cm
(*Sinhala* Polon Muhudu-naya; *Tamil* Kadal Pambu

DESCRIPTION Body medium sized. Tail flattened and paddle-like. Head short and somewhat triangular. Neck distinct. Ventral scales in anterior half of body broad and larger than body scales; on rest of body they gradually become narrower. Body dorsally grey in colour, and belly region paler or whitish in most individuals. Some may bear 25–35 dark spots on dorsal surface. **DISTRIBUTION** Coasts of northern region of Sri Lanka. Extra-limital: coastal waters of Arabian Gulf, to eastern China and Southeast Asia through South Asia. **HABITAT AND HABITS** Lives 5–10km offshore in northern region of Sri Lanka. Feeds on various types of fish. Ovoviviparous, giving birth to about 3–6 young. Highly venomous and becomes aggressive when caught in fishing nets.

Plain

Ventral

Bar patterned

Slender Sea Snake ■ *Microcephalophis gracilis* 100cm
(*Sinhala* Kudahis Muhudu-naya *Tamil* Kadal Pambu)

DESCRIPTION Body medium sized and elongate; slender anteriorly and compressed posteriorly. Tail flattened and paddle-like. Head small and indistinct from neck; snout projects beyond lower jaw. Body scales hexagonal and bear 2–3 tubercles or prominent keel. Ventral scales hexagonal and divided by longitudinal groove. Adults greyish above and paler below, with indistinct spots or bands in anterior region of body. **DISTRIBUTION** Coastal regions all over Sri Lanka. Extra-limital: coasts of Arabian Gulf to southern China and New Guinea, through South and Southeast Asia. **HABITAT AND HABITS** Inhabits shallow coastal waters. Feeds on benthic eels. Ovoviviparous, giving birth to 1–6 young. Inoffensive snake that is occasionally caught in fishing nets in Sri Lanka. Highly venomous.

Hexagonal ventral scales with longitudinal grooves

Dorso-lateral

HOMALOPSIDAE (INDO-AUSTRALIAN MUD SNAKES)

The Indo-Australian mud snakes are a group of snakes that show partial to strong aquatic affinities. Many species live in freshwater habitats like lakes, marshes, swamps, streams and paddy fields, while others occupy brackish water habitats such as estuaries, mangroves and other shallow-water coastal areas. Most have valved nostrils and dorsally directed eyes situated at the top of the head as adaptations to aquatic life. They also have enlarged fangs at the back of the mouth. They are ovoviviparous and give birth to live young. Fifty-three species in 27 genera are known from South Asia, Southeast Asia and Australasia. Sri Lanka is home to three species in three genera. The Rainbow Water Snake *Enhydris enhydris* is known from only a single specimen in Sri Lanka.

Dogface Water Snake ▪ *Cerberus rynchops* 45cm
(*Sinhala* Kunudiya Kaluwa; *Tamil* Uppu-eri Pambu)

DESCRIPTION Body medium sized. Tail short and slightly compressed. Head long and distinct from neck; snout broadly rounded; resembles a dog's face; upper jaw projects outwards. Eyes small and beady with rounded pupils. Scales dull and strongly keeled; costals 21–25; ventrals 135–150; subcaudals 61–70. Dorsum dark grey; occasionally light brown

with faint dark blotches and dark line along sides of head, across eyes. Belly yellowish-cream with dark grey areas. **DISTRIBUTION** Coastal regions. Extra-limital: Bangladesh, India, Pakistan, Andaman and Nicobar islands, and west coast of Thailand. **HABITAT AND HABITS** Occupies estuaries, lagoons, mangroves and mudflats. Nocturnal and aquatic. During daytime, hides in crab holes, emerging at night to feed. Diet comprises fish (such as mudskippers and gobies), crabs and frogs. Ovoviviparous, giving birth to 6–30 young. Mildly venomous inoffensive snake that rarely bites when handled.

Dorso-lateral

Dog's face head

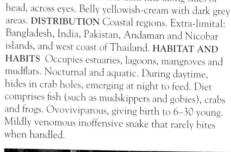

Ventral patterned with broad spots

Gerard's Water Snake ◼ *Gerarda prevostiana* 53cm
(*Sinhala* Prevostige Diya Bariya)

DESCRIPTION Body small and relatively long. Tail short. Head distinct from neck. Eyes small with vertically elliptical pupils. Scales smooth. Dorsum unpatterned grey or brown; lips and lower scales of dorsum cream; belly brownish-cream with median dark streaks.
DISTRIBUTION Known from a few localities on western and north-western coast. Extra-limital: India to the Philippines through Bangladesh, Myanmar, Thailand, Singapore,

Malaysia, Borneo and Cambodia. **HABITAT AND HABITS** Lives in brackish water habitats including mangroves, swamps and river mouths. Nocturnal and aquatic. Rare species known from only a few specimens. Known to hide in mud lobster mounds during day. Feeds mainly on soft-shelled crabs, though also eats fish and shrimps. Mildly venomous.

Grey colouration

Brown colour morph with spots

VIPERIDAE (VIPERS AND PIT VIPERS)

All snakes of the Viperidae family are venomous, and the group includes about 230 species worldwide. The body of a viperid is short and bulky, with a distinct or slightly triangular head bearing small scales. The scales on the body and head are heavily keeled, and thus are dull and lacklustre in appearance. Large, movable fangs are situated in the front of the mouth, and are folded back against the palate when not in use. The Viperidae family consists of two main groups: 'true vipers' (Viperinae) and 'pit vipers' (Crotalinae). The pit vipers bear a specialized heat-sensitive pit (loreal pit) between the nostril and eye, which allows them to detect warm-blooded prey in the dark. Members of the family feed mainly on small mammals, frogs, birds, skinks and in some cases invertebrates. They are mostly ovoviviparous, giving birth to live young. In Sri Lanka, the family comprises six species in the two groups, vipers and pit vipers, with three species being endemic to Sri Lanka. Another species (*Hypnale* sp.) has recently been discovered, but awaits formal description.

Russell's Viper ▪ *Daboia russelii* 130cm
(*Sinhala* Tith Polonga; *Tamil* Kannadi Viriyan)

DESCRIPTION Triangular head distinct from neck, with conspicuous 'V'-shaped mark. Body robust. Tail thin and short. Dorsal body colour light brownish-grey with keeled dull scales. Dorsally there are 20–30 large, ovate spots in row along vertebral line. Sides have series of spots that are reddish-brown inside with black outer edges. Ventral side white with slight tinge of pink, with brown spots. **DISTRIBUTION** Widely distributed from sea level up to about 1,500m above. Common in dry zone. Extra-limital: Pakistan, India, Nepal and Bangladesh. **HABITAT AND HABITS** Main habitats are grasslands, scrub jungles, plantations (tea, coffee, coconut, rubber), vicinity of paddy fields, and *chena* (slash-and-burn) cultivation areas. Nocturnal and terrestrial. Occasionally may cross roads during day. Feeds mostly on rodents. Ovoviviparous, producing 5–50 young Highly venomous. Hisses loudly when cornered or disturbed.

Dorso-lateral

Close-up of head

Saw Scaled Viper ■ *Echis carinatus* 45cm
(*Sinhala* Weli Polnaga; *Tamil* Surutai Vireyan)

DESCRIPTION Body small and stout. Neck distinct. Head ovoid with large eyes and vertical pupils. Distinct cruciform or trident white mark on head. Body cylindrical and rough due to serrated keeled scales. Dorsal side greyish, reddish, olive or pale brown, with distinct whitish pattern and series of dark brown patches. Laterally, there is a pale continuous pattern. Belly white with brown spots. **DISTRIBUTION** Mainly in dry coastal regions (semi-arid zone) of north and north-west (for example Mannar, Delft island, Panama, Bundala and Jaffna). Extra-limital: Middle East, Pakistan and India. **HABITAT AND HABITS** Favours sand dunes but also found under logs and stones during day. Nocturnal and terrestrial. Feeds on scorpions, centipedes, insects, small lizards and mice. Ovoviviparous, giving birth to 3–15 young. When threatened, makes hissing sound by rubbing body scales together. Very aggressive and readily bites without hesitation. Highly venomous.

Dorso-lateral

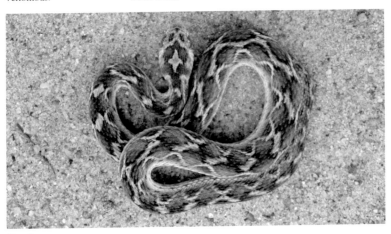

Dorsal

Merrem's Hump-nosed Viper ■ *Hypnale hypnale* 40cm
(*Sinhala* Polon Thelissa; *Tamil* Kopi Viriyan)

DESCRIPTION Body small and stout. Head flat and triangular, with snout pointed upwards. Loreal pit between nostril and eye. Body dull in colour, varying from light brown, olive or grey, to dark brown with darker scattered lateral spots along it. Neck region lighter, with small spots. Tip of tail lighter. Belly pale, mottled with minute brown or grey spots. **DISTRIBUTION** Coastal lowlands to 800m above sea level. Common in wet zone. Extra-limital: Western Ghats of India. **HABITAT AND HABITS** Found in home gardens, plantations, grassland and undisturbed forests. Nocturnal and terrestrial, occasionally ascending trees. During day, hides under logs and among leaf litter, which nicely camouflage it. When resting, keeps heads at 45-degree angle to the ground. Feeds on lizards, frogs and mice. Ovoviviparous, giving birth to 3–18 young. Easily agitated and bites viciously when provoked. Highly venomous.

Dorso-lateral, grey

Close-up of head

Dorso-lateral, light brown

Millard's Hump-nosed Viper ■ *Hypnale nepa* 30cm **e**
(*Sinhala* Kandukara Mukalan Thelissa)

DESCRIPTION Body small and stout. Head flattened and triangular. Snout pointed upwards with tip that bears small protuberance. Loreal pit between nostril and eye. Body covered with feebly keeled scales. Body colour varies from yellowish-tan to dark brown, with 2 rows of distinct suboval or subtriangular blotches that meet on dorsal midline.

Dark stripe runs across eye to cheek. Belly much lighter than body. **DISTRIBUTION** Central Highlands at 1,250–1,850m above sea level. Endemic to Sri Lanka. **HABITAT AND HABITS** Occurs in primary and secondary forests, as well as forest edges. Nocturnal and terrestrial. During day hides under logs, stones and decaying leaf litter. When resting, keeps head at 45-degree angle to the ground. Feeds on lizards and small snakes. Ovoviviparous, giving birth to 5–8 young. Moderately venomous.

Close-up of head

Dorso-lateral

Lowland Hump-nosed Viper ■ *Hypnale zara* 35cm ⓔ
(*Sinhala* Mukalan Thelissa)

DESCRIPTION Body small and stout. Head flattened and triangular. Tip of snout distinctly raised compared to snouts of other hump-nosed vipers, and bears large protuberance (larger than that of Millard's Hump-nosed Viper, see p. 161). Loreal pit between nostril and eye. Body colour dull, and varies from yellowish-brown to dark brown, with two rows of distinct suboval or subtriangular blotches that meet on dorsal midline. Dark stripe runs across eye to cheek. Belly much lighter than upper parts. **DISTRIBUTION** Lowland wet zone at 60–1,000m above sea level. Endemic to Sri Lanka. **HABITAT AND HABITS** Found in primary and secondary forests, and occasionally home gardens. Nocturnal and terrestrial. Occasionally climbs short shrubs and bushes. During day, hides under logs, stones and decaying leaf litter. When resting, keeps head at 45-degree angle to the ground. Feeds on mice and small reptiles. Ovoviviparous, giving birth to 5–8 young. Moderately venomous.

Dorso-lateral, yellowish-brown morph

Dorso-lateral, dark brown morph *Close-up of head*

Sri Lanka Green Pit Viper ■ *Trimeresurus trigonocephalus* 100cm ⓔ
(*Sinhala* Pala Polonga; *Tamil* Patchchai Viriyan)

DESCRIPTION Body stout, medium sized and thick. Head large and triangular. Neck distinct. Loreal pit is present. Tail short and prehensile. Body light greenish with black variegated pattern on dorsal side (absent in some individuals); scales dull in appearance. Black temporal line on either side of head and tail. Posterior part of tail is black. Ventral aspect light greenish-yellow or grey in some individuals. **DISTRIBUTION** Widely distributed in three main climatic zones (wet, dry and intermediate) up to 1,200m above sea level. Endemic to Sri Lanka. **HABITAT AND HABITS** Common in lowland rainforests and submontane forests, usually seen among shrubs along streams. Nocturnal and arboreal, and may descend to the ground in search of food. Often found on low shrubs rather than on tall trees. Feeds on frogs, rodents and birds. Ovoviviparous, giving birth to 5–26 young. When agitated, folds forepart of body into sinuous loops and lashes out rapidly to bite any object encountered. Moderately venomous.

Dorso-lateral, plain colouration

Close-up of head

Dorso-lateral aspect of typical patterned colouration

Jan's Blind Snake ▪ *Gerrhopilus mirus* 14cm ⓔ
(*Sinhala* Heen Kanaullla)

DESCRIPTION Body small, with same girth from head to tail. Tail short with blunt-ended tip. Eyes minute, each covered with transparent scale. Numerous distinct sebaceous glands cover head shields. Scales: nasal completely divided; 18 scales around mid-body; 330–360 transverse scale rows. Dorsal and lateral head profiles rounded. Ground colour of dorsal aspect dark brown, head lighter, while ventral side is paler in colour. **DISTRIBUTION** Wet and intermediate zones from sea level to 1,000m above. Endemic to Sri Lanka. **HABITAT AND HABITS** Nocturnal and fossorial. Frequents soil and decaying leaf litter. Feeds on small insect larvae, pupae and adult insects. Oviparous and lays eggs. Known to secrete foul-smelling substance when captured. Non-venomous.

Dorsal

Common Blind Snake ■ *Indotyphlops braminus* 15cm
(*Sinhala* Dumbutu Kanaulla; *Tamil* Seer Pambu)

DESCRIPTION Body small, with same girth from head to tail. Tail short, ending in sharp distinct spine. Eyes minute, each covered with transparent scale. Dorsal and lateral head profiles rounded. Scales: costals 20 around mid-body; 290–330 transverse scale rows. Visible sebaceous glands underneath head scales, in epidermis. Ground colour of dorsal aspect dark brown to black, while ventral side is paler in colour. **DISTRIBUTION** Widely distributed in wet, intermediate and dry zones from sea level to 1,000m above. Extra-limital: Africa, the Mediterranean, Middle East, South Asia, Southeast Asia, New Guinea, Polynesian islands, Micronesia, and Central and South America. **HABITAT AND HABITS** Nocturnal and fossorial. Frequents sandy soil and decaying leaf litter. Feeds on small insect larvae, pupae and adult insects. Oviparous, laying 2–7 eggs. Known to poke with caudal spines when captured, and simultaneously secrete a foul-smelling substance. Non-venomous.

Ventral

Dorso-lateral

CROCODYLIDAE (CROCODILES)

The Crocodilia order comprises large, solidly built, lizard-like reptiles with long, flattened snouts, laterally compressed tails, and eyes, ears and nostrils at the top of the head. Their evolutionary beginnings date back to the early Triassic period more than 250 million years ago. They are aquatic, and are the world's largest living reptiles, with some individuals reaching 6m in length and weighing almost a tonne. In Sri Lanka, the Crocodilia are represented by one family, the Crocodylidae. This family comprises 15 extant species distributed in the tropical regions of Africa, Asia, Australasia and the Americas. All species are predators of small to large prey. Two species occur in Sri Lanka, and some large individuals may pose a danger to humans and livestock. All species are linked to wetlands such as rivers, lakes, dams and mangroves. They are oviparous, laying large, soft-shelled eggs in mounds or holes dug in the ground.

Mugger Crocodile ■ *Crocodylus palustris* 4.5m
(*Sinhala* Häla Kimbula; *Tamil* Kulathi Muthalai)

DESCRIPTION Body large and elongated. Snout broad. Well-developed shields behind head and neck. Feet webbed. Juveniles light tan or brown, with dark cross-bands on body and tail. Adults grey to brown. **DISTRIBUTION** Widely distributed in all water sources in dry zone lowlands. Common in Yala National Park and Jaffna peninsula. Extra-limital: India, eastern Iran, Pakistan and Nepal. **HABITAT AND HABITS** Found in various aquatic habitats such as rivers, estuaries, lagoons, tanks, canals, agro-wells and other waterways. Mainly nocturnal in behaviour, but diurnal activities such as basking, foraging, migrating over land, and digging nests and burrows are also seen. In some lakes 50–60 muggers can be seen basking. Hatchlings feed on insects, amphibians, fish and crustaceans; adults feed on fish, terrapins, tortoises, lizards, snakes, birds, monkeys and dogs. Also attacks large mammals and occasionally humans. Oviparous, laying 10–30 eggs in pit dug on sandy bank close to water.

Adult

Juvenile

Saltwater Crocodile ■ *Crocodylus porosus* 6.2m

(*Sinhala* Geta Kimbula; *Tamil* Semmukhan, Semmukku Muthalei)

DESCRIPTION Head large. Snout more elongated than in Mugger Crocodile (see opposite), with pair of ridges running from orbit to centre of snout. Neck more granular than in Mugger. Juveniles brightly coloured – black spotted or blotched on pale yellow or grey background. Colouration of dorsum less bright in adults. **DISTRIBUTION** Locally known from Chillaw to down south towards Matara in coastal rivers, marshes, streams and north-eastern coastal region. Extra-limital: India to Fiji through South Asia, Southeast Asia, the Philippines and Australasia. **HABITAT AND HABITS** Occurs in rivers that drain into the sea, mainly on eastern, western and southern coasts with riverine mangrove plants and streams. Aquatic, and nocturnal with diurnal activities. Adult males usually solitary; emit roars and bellows. Hatchlings feed on arthropods, crustaceans, fish and frogs; adults feed on fish, chelonians, birds and mammals, occasionally including humans. Oviparous; female constructs mound nest with decaying leaves, in which 20–50 eggs are deposited.

Juvenile

Adult

(Follows latest nomenclature available as of December 2016)

Abbreviations of IUCN status

CR Critically Endangered
EN Endangered
VU Vulnerable
NT Near Threatened
LC Least Concern
DD Data Deficient

Scientific Name	Common Name	Status	IUCN Status
Emydidae (Freshwater Terrapins)			
Trachemys scripta	Red-eared Slider	Introduced	NE
Testudinidae (Land Tortoises)			
Geochelone elegans	Indian Star Tortoise	Indigenous	NT
Geoemydidae (Freshwater Terrapins and Turtles)			
Melanochelys trijuga	Parker's Black Turtle	Indigenous	LC
Trionychidae (Softshell Turtles and Terrapins)			
Lissemys ceylonensis	Sri Lanka Flapshell Turtle	Endemic	LC
Cheloniidae (Sea Turtles)			
Caretta caretta	Loggerhead Sea Turtle	Indigenous	EN
Chelonia mydas	Green Turtle	Indigenous	EN
Eretmochelys imbricata	Hawksbill Sea Turtle	Indigenous	EN
Lepidochelys olivacea	Olive Ridley Sea Turtle	Indigenous	EN
Dermochelyidae (Leatherback Sea Turtles)			
Dermochelys coriacea	Leatherback Sea Turtle	Indigenous	CR
Agamidae (Dragon Lizards)			
Calotes calotes	Green Garden Lizard	Indigenous	LC
Calotes ceylonensis	Painted Lip Lizard	Endemic	NT
Calotes desilvai	Desilvas' Whistling Lizard	Endemic	CR
Calotes liocephalus	Crestless Lizard	Endemic	CR
Calotes liolepis	Sri Lanka Whistling Lizard	Endemic	NT
Calotes manamendrai	Manamendra's Whistling Lizard	Endemic	NE
Calotes nigrilabris	Black Cheek Lizard	Endemic	EN
Calotes pethiyagodai	Pethiyagoda's Crestless Lizard	Endemic	EN
Calotes versicolor	Common Garden Lizard	Indigenous	LC
Ceratophora aspera	Rough Horn Lizard	Endemic	EN
Ceratophora erdeleni	Erdelen's Horn Lizard	Endemic	CR
Ceratophora karu	Karunaratne's Horn Lizard	Endemic	CR
Ceratophora stoddartii	Rhinohorn Lizard	Endemic	EN
Ceratophora tennentii	Leafnose Lizard	Endemic	CR
Cophotis ceylanica	Pygmy Lizard	Endemic	EN
Cophotis dumbara	Knuckles Pygmy Lizard	Endemic	CR
Lyriocephalus scutatus	Lyre Head Lizard	Endemic	VU
Otocryptis nigristigma	Black Spotted Kangaroo Lizard	Endemic	LC
Otocryptis wiegmanni	Sri Lankan Kangaroo Lizard	Endemic	LC
Sitana bahiri	Bahir's Fan-throat Lizard	Endemic	NE
Sitana devakai	Devaka's Fan-throat Lizard	Endemic	NE

Scientific Name	Common Name	Status	IUCN Status
Chameleonidae (Chameleons)			
Chamaeleo zeylanicus	Sri Lankan Chameleon	Indigenous	EN
Gekkonidae (Geckoes)			
Calodactylodes illingworthorum	Lankan Golden Gecko	Endemic	EN
Cnemaspis alwisi	Alwis' Day Gecko	Endemic	NT
Cnemaspis amith	Amith's Day Gecko	Endemic	CR
Cnemaspis clivicola	Montain Day Gecko	Endemic	CR
Cnemaspis gemunu	Gemunu's Day Gecko	Endemic	CR
Cnemaspis kallima	Gammaduwa Day Gecko	Endemic	CR
Cnemaspis kandiana	Kandyan Day Gecko	Endemic	EN
Cnemaspis kumarasinghei	Kumarasinghe's Day Gecko	Endemic	EN
Cnemaspis latha	Elegant Day Gecko	Endemic	CR
Cnemaspis menikay	Jewel Day Gecko	Endemic	CR
Cnemaspis molligodai	Molligod's Day Gecko	Endemic	EN
Cnemaspis pava	Little Day Gecko	Endemic	CR
Cnemaspis phillipsi	Phillip's Day Gecko	Endemic	CR
Cnemaspis podihuna	Dwarf Day Gecko	Endemic	VU
Cnemaspis pulchra	Rakvana Day Gecko	Endemic	CR
Cnemaspis punctata	Dotted Day Gecko	Endemic	CR
Cnemaspis rajakarunai	Rajakaruna's Day Gecko	Endemic	NE
Cnemaspis rammalensis	Rammale Day Gecko	Endemic	NE
Cnemaspis retigalensis	Ritigala Day Gecko	Endemic	CR
Cnemaspis samanalensis	Peakwilderness Day Gecko	Endemic	CR
Cnemaspis scalpensis	Gannoruva Day Gecko	Endemic	EN
Cnemaspis silvula	Forest Day Gecko	Endemic	NE
Cnemaspis tropidogaster	Roughbelly Day Gecko	Endemic	DD
Cnemaspis upendrai	Upendra's Day Gecko	Endemic	CR
Hemidactylus platyurus	Frill-tail Gecko	Indigenous	DD
Cyrtodactylus cracens	Narrow-headed Forest Gecko	Endemic	EN
Cyrtodactylus edwardtaylori	Taylor's Forest Gecko	Endemic	CR
Cyrtodactylus fraenatus	Great Forest Gecko	Endemic	CR
Cyrtodactylus ramboda	Ramboda Forest Gecko	Endemic	CR
Cyrtodactylus soba	Knuckles Forest Gecko	Endemic	CR
Cyrtodactylus subsolanus	Rakwana Forest Gecko	Endemic	CR
Cyrtodactylus trieda	Spotted Bowfinger Gecko	Endemic	VU
Cyrtodactylus collegalensis	Collegal Rock Gecko	Indigenous	DD
Cyrtodactylus yakhuna	Blotch Bowfinger Gecko	Endemic	VU
Gehyra mutilata	Four-claw Gecko	Indigenous	LC
Hemidactylus depressus	Kandyan Gecko	Endemic	LC
Hemidactylus frenatus	Common House-gecko	Indigenous	LC
Hemidactylus hunae	Spotted Giant-gecko	Endemic	EN
Hemidactylus lankae	Termite Hill Gecko	Endemic	LC
Hemidactylus leschenaultii	Bark Gecko	Indigenous	LC
Hemidactylus parvimaculatus	Spotted House-gecko	Indigenous	LC
Hemidactylus pieresii	Pieres' Gecko	Endemic	EN
Hemidactylus scabriceps	Scaly Gecko	Indigenous	DD
Hemiphyllodactylus typus	Slender Gecko	Indigenous	VU
Lepidodactylus lugubris	Indo-Pacific Mourning Gecko	Indigenous	VU
Lacertidae (Wall Lizards)			
Ophisops leschenaultii	Leschenault's Snake-eye Lizard	Indigenous	CR
Ophisops minor	Lesser Snake-eye Lizard	Indigenous	CR
Lygosomidae (Skinks)			
Lygosoma punctatus	Dotted Skink	Indigenous	LC
Lygosoma singha	Taylor's Skink	Endemic	DD

Scientific Name	Common Name	Status	IUCN Status
Mabuyidae (Skinks)			
Dasia halianus	Haly's Tree Skink	Endemic	NT
Eutropis austini	Austin's Skink	Endemic	NE
Eutropis beddomii	Beddome's Striped Skink	Indigenous	EN
Eutropis bibronii	Bibron's Sand Skink	Indigenous	EN
Eutropis carinata	Common Skink	Indigenous	LC
Eutropis floweri	Taylor's Skink	Endemic	EN
Eutropis greeri	Greer's Skink	Endemic	NE
Eutropis madaraszi	Spotted Skink	Endemic	VU
Eutropis tammanna	Tmmanna Skink	Endemic	LC
Ristellidae (Skinks)			
Lankascincus deignani	Deignan's Lanka Skink	Endemic	EN
Lankascincus deraniyagalae	Deraniyagal's Lanka Skink	Endemic	EN
Lankascincus dorsicatenatus	Catenated Litter Skink	Endemic	EN
Lankascincus fallax	Common Lanka Skink	Endemic	LC
Lankascincus gansi	Gans' Lanka Skink	Endemic	VU
Lankascincus greeri	Geer's Lanka Skink	Endemic	EN
Lankascincus munindradasai	Munidradasa's Lanka Skink	Endemic	CR
Lankascincus sripadensis	Peak Wilderness Lanka Skink	Endemic	CR
Lankascincus taprobanensis	Smooth Lanka Skink	Endemic	EN
Lankascincus taylori	Taylor's Lanka Skink	Endemic	EN
Scincidae (Skinks)			
Chalcides cf. *ocellatus*	White-spotted Skink	Endemic	DD
Chalcidoseps thwaitesii	Four-toe Snake Skink	Endemic	CR
Nessia bipes	Smith's Snake Skink	Endemic	EN
Nessia burtonii	Three-toe Snake Skink	Endemic	LC
Nessia deraniyagalai	Deraniyagala's Snake Skink	Endemic	DD
Nessia didactylus	Two-toe Snake Skink	Endemic	EN
Nessia hickanala	Sharkhead Snake Skink	Endemic	CR
Nessia layardi	Layard's Snake Skink	Endemic	EN
Nessia monodactylus	Toeless Snake Skink	Endemic	EN
Nessia sarasinorum	Sarasin's Snake Skink	Endemic	VU
Sphenomorphidae (Skinks)			
Sphenomorphus dussumieri	Dussumier's Litter Skink	Indigenous	DD
Sphenomorphus megalops	Annandale's Litter Skink	Endemic	DD
Varanidae (Monitors)			
Varanus bengalensis	Land Monitor	Indigenous	LC
Varanus salvator	Water Monitor	Indigenous	LC
Acrochordidae (File Snakes)			
Acrochordus granulatus	Wart Snake	Indigenous	VU
Cylindrophiidae (Pipe Snakes)			
Cylindrophis maculata	Sri Lanka Pipe Snake	Endemic	NT
Uropeltidae (Shield-tailed Snakes)			
Platyplectrurus madurensis	Madura Blind Snake	Indigenous	DD
Rhinophis blythii	Blyth's Earth Snake	Endemic	EN
Rhinophis dorsimaculatus	Orange Shield Tail	Endemic	DD
Rhinophis drummondhayi	Drummond-hay's Earth Snake	Endemic	EN
Rhinophis erangaviraji	Eranga Viraj's Earth Snake	Endemic	CR
Rhinophis homolepis	Kelaarts Earth Snake	Endemic	EN
Rhinophis lineatus	Lineate Earth Snake	Endemic	CR
Rhinophis melanogaster	Black Shield-tail	Endemic	EN
Rhinophis oxyrynchus	Schneider's Earth Snake	Endemic	LC
Rhinophis philippinus	Cuvier's Earth Snake	Endemic	EN
Rhinophis phillipsi	Phillips' Shield-tail	Endemic	EN

Scientific Name	Common Name	Status	IUCN Status
Rhinophis porrectus	Willey's Earth Snake	Endemic	DD
Rhinophis punctatus	Muller's Earth Snake	Endemic	DD
Rhinophis ruhunae	Ruhunu Shield-tail	Endemic	DD
Rhinophis saffragamus	Large Shield-tail	Endemic	EN
Rhinophis tricoloratus	Deraniyagala's Shield-tail	Endemic	CR
Rhinophis zigzag	Zigzag Earth Snake	Endemic	CR
Pythonidae (Pythons)			
Python molurus	Indian Python	Indigenous	LC
Boidae (Boas)			
Eryx conicus	Rough-scaled Sand Boa	Indigenous	VU
Colubridae (Typical Snakes)			
Ahaetulla nasuta	Green Vine Snake	Indigenous	LC
Ahaetulla pulverulenta	Brown Vine Snake	Indigenous	LC
Amphiesma stolatum	Buff Striped Keelback	Indigenous	LC
Aspidura brachyorrhos	Boie's Roughside	Endemic	VU
Aspidura ceylonensis	Black Spine Snake	Endemic	EN
Aspidura copei	Cope's Roughside	Endemic	DD
Aspidura deraniyagalae	Deraniyagala's Roughside	Endemic	CR
Aspidura drummondhayi	Drummond-Hay's Roughside	Endemic	EN
Aspidura guentheri	Ferguson's Roughside	Endemic	NT
Aspidura trachyprocta	Common Roughside	Endemic	EN
Atretium schistosum	Olive Keelback	Indigenous	LC
Argyrogena fasciolata	Banded Racer	Indigenous	DD
Balanophis ceylonensis	Sri Lanka Keelback	Endemic	EN
Boiga barnesii	Barnes' Cat Snake	Endemic	VU
Boiga beddomei	Beddoms Cat Snake	Indigenous	NT
Boiga ceylonensis	Sri Lanka Cat Snake	Indigenous	LC
Boiga forsteni	Forsten's Cat Snake	Indigenous	NT
Boiga trigonatus	Gamma Cat Snake	Indigenous	LC
Chrysopelea ornata	Ornate Flying Snake	Indigenous	VU
Chrysopelea taprobanica	Striped Flying Snake	Endemic	LC
Coeloganthus helena	Trinket Snake	Indigenous	LC
Dendrelaphis bifrenalis	Boulenger's Bronzeback	Indigenous	NT
Dendrelaphis caudolineolatus	Striped Bronzeback	Indigenous	VU
Dendrelaphis oliveri	Oliver's Bronzeback	Endemic	DD
Dendrelaphis schokari	Common Bronzeback	Endemic	LC
Dendrelaphis sinharajensis	Sinharaja Bronzeback	Endemic	NE
Dendrelaphis tristis	Front-spot Bronzeback	Indigenous	LC
Dryocalamus gracilis	Scarce Bridal Snake	Indigenous	DD
Dryocalamus nympha	Bridal Snake	Indigenous	LC
Liopeltis calamaria	Reed Snake	Indigenous	NT
Lycodon carinata	Sri Lanka Wolf Snake	Endemic	EN
Lycodon aulicus	Wolf Snake, House Snake	Indigenous	LC
Lycodon osmanhilli	Flowery Wolf Snake	Endemic	LC
Lycodon striatus	Shaw's Wolf Snake	Indigenous	LC
Macropisthodon plumbicolor	Green Keelback,	Indigenous	NT
Oligodon arnensis	Common Kukri Snake/ Banded Kukri	Indigenous	LC
Oligodon calamarius	Templeton's Kukri Snake	Endemic	EN
Oligodon sublineatus	Dumerul's Kukri Snake	Endemic	LC
Oligodon taeniolata	Variegated Kukri Snake	Indigenous	LC
Ptyas mucosa	Rat Snake	Indigenous	LC
Sibynophis subpunctatus	Jerdon's Polyodent	Indigenous	NT
Xenochrophis asperrimus	Checkered Keelback	Endemic	LC
Xenochrophis cf. piscator	Checkered Keelback	Endemic	LC

Scientific Name	Common Name	Status	IUCN Status
Elapidae (Cobras, Coral Snakes, Kraits, Mambas and Sea Snakes)			
Bungarus caeruleus	Common Krait	Indigenous	LC
Bungarus ceylonicus	Sri Lanka Krait	Endemic	VU
Calliophis haematoetron	Bloody-vented Coral Snake	Endemic	CR
Calliophis melanurus	Sri Lanka Coral Snake	Indigenous	NT
Naja naja	Indian Cobra	Indigenous	LC
Hydrophis bituberculatus	Peter's Sea Snake	Indigenous	DD
Hydrophis curtus	Shaw's Sea Snake	Indigenous	LC
Hydrophis cyanocinctus	Annulated Sea Snake	Indigenous	LC
Hydrophis fasciatus	Striped Sea Snake	Indigenous	LC
Hydrophis jerdonii	Jerdon's Sea Snake	Indigenous	LC
Hydrophis lapemoides	Persian Gulf Sea Snake	Indigenous	LC
Hydrophis mammilaris	Bombay Gulf Sea Snake	Indigenous	DD
Hydrophis ornatus	Gray's Sea Snake	Indigenous	LC
Hydrophis schistosa	Hook-nosed Sea Snake	Indigenous	LC
Hydrophis spiralis	Narrow-banded Sea Snake	Indigenous	LC
Hydrophis stokesii	Stoke's Sea Snake	Indigenous	LC
Hydrophis stricticollis	Guenther's Sea Snake	Indigenous	DD
Hydrophis platurus	Yellow-belly Sea Snake	Indigenous	LC
Hydrophis viperinus	Viperine Sea Snake	Indigenous	LC
Microcephalophis gracilis	John's Sea Snake	Indigenous	LC
Homalopsidae (Indo-Australian Mud Snakes)			
Cerberus rynchops	Dog-faced Water Snake	Indigenous	LC
Gerarda prevostianus	Gerard's Water Snake	Indigenous	EN
Enhydris enhydris	Rainbow Mud Snake	Indigenous	NE
Gerrhopilidae (Blind Snakes)			
Gerrhopilus ceylonicus	Smith's Blind Snake	Endemic	DD
Gerrhopilus lankaensis	Lanka Blind Snake	Endemic	CR
Gerrhopilus leucomelas	Pied Gerrhopilus	Endemic	CR
Gerrhopilus malcolmi	Malcolm's Blind Snake	Endemic	DD
Gerrhopilus mirus	Jan's Blind Snake	Endemic	CR
Gerrhopilus porrectus	Stoliczka's Blind Snake	Indigenous	EN
Gerrhopilus tenebrarum	Taylor's Blind Snake	Endemic	DD
Gerrhopilus veddae	Veddha's Blind Snake	Endemic	DD
Gerrhopilus violaceus	Violet Blind Snake	Endemic	DD
Viperidae (Vipers and Pit Vipers)			
Daboia russelii	Russell's Viper	Indigenous	LC
Echis carinatus	Saw Scale Viper	Indigenous	VU
Hypnale hypnale	Merrem's Hump-nosed Viper	Indigenous	LC
Hypnale nepa	Merrem's Hump-nosed Viper	Endemic	EN
Hypnale zara	Zara's Hump-nosed Viper	Endemic	VU
Hypnale 'amal'	Amal's Hump-nosed Viper	Endemic	CR
Trimeresurus trigonocephalus	Sri Lanka Green Pit Viper	Endemic	LC
Typhlopidae (Blind Snakes)			
Indotyphlops braminus	Common Blind Snake	Indigenous	LC
Crocodylidae (Crocodiles)			
Crocodylus palustris	Mugger Crocodile	Indigenous	NT
Crocodylus porosus	Saltwater Crocodile	Indigenous	EN

Further Reading

Das, I. & de Silva, Anslem (2011) *A Photographic Guide to Snakes and Other Reptiles of Sri Lanka*. New Holland, UK.

Deraniyagala, P. E. P. (1953) *A Coloured Atlas of Some Vertebrates from Ceylon – Tetrapod Reptilia*, Vol. 2, National Museums of Sri Lanka. The Ceylon Government Press. Colombo, Sri Lanka.

Deraniyagala, P. E. P. (1955) *A Coloured Atlas of Some Vertebrates from Ceylon – Serpentoid Reptilia*, Vol. 3, National Museums of Sri Lanka. The Ceylon Government Press. Colombo, Sri Lanka.

de Silva, Anslem (1990) *Colour Guide to the Snakes of Sri Lanka*. R. & A. Publishing Ltd. Avon, England.

de Silva, P. H. D. H. (1980) *Snake Fauna of Sri Lanka: With Special Reference to Skull, Dentition and Venom in Snakes*. National Museums of Sri Lanka, Colombo, Sri Lanka.

Smith, M. A. (1943) *Fauna of British India, Ceylon and Burma – Reptilia and Amphibia*, Vol. III – Serpentes. Taylor and Francis, London, England.

Somaweera, R. (2006) *Sri Lankawe Sarpayin* ('The Snakes of Sri Lanka'). Wildlife Heritage Trust of Sri Lanka, Colombo.

Somaweera, R. & Somaweera, N. (2009) *Lizards of Sri Lanka: A Colour Guide with Field Keys*. Edition Chimaria, Frankfurt, Germany.

Wall, F. (1921) *Ophidia Taprobanica or the Snakes of Ceylon*. H. R. Cottle, Govt printer, Colombo, Sri Lanka.

Acknowledgements

We thank Sameera Karunarathne, Udaya Chanaka, Thushan Kapurusinghe, Suraj Goonewardene, Luxshmanan Nadaraja, Usui Toshikazu, Mike Anthonisz and the late Carl Gans for providing excellent photographs. We highly appreciate Gayathri Selvaraj, Sameera Karunarathne and Ruchira Somaweera for their valuable comments on the drafts of this book. Our colleauges Panduka de Silva, Suraj Goonawardena, Imesh Nuwan Bandara, Tharaka Priyadarshane, Ishara Wijewardehane, Senani Karunarthne, Krishantha Sameera de Zoysa, Duminda Dissanayake and Rajnish Vandercone are sincerely thanked for accompanying us in the field on many occasions. We are very grateful to Krishantha Sameera de Zoysa for preparing the map on climatic zones on the inside back cover. The Department of Wildlife Conservation and the Department of Forest Conservation of Sri Lanka are thanked for research permits issued for various reptile-research projects by the two authors. We gratefully acknowledge the Mohamed Bin Zeyed Species Conservation Fund for the generous grants to AdS for sea snake, crocodile and skink-research projects. We thank John Beaufoy Publishing Ltd, Oxford, UK for accepting and publishing this guide. Finally, we lovingly remember our families for tolerating numerous reptiles in our homes, and bearing long periods of absence when we are out in the field.

■ INDEX ■